From Buffer to Boon: How Floodplain Reconnection Benefits Rivers and Beyond

Jessy

Table of Contents

Chapter 1: Introduction...12
    Stream Restoration Practices..12
    Controversy Surrounding Reconnection...13
    Current Floodplain Reconnection Projects 14
    Literature Review – Biological Responses to Floodplain Reconnection and Nutrient
    Enrichment 15
        Periphyton..15
        Macroinvertebrates 17
        Fish..18
        Trophic Level Interactions 19
        Abiotic Factors..20
        River Continuum Concept...20
    Big Spring Run – Case Study 21
        Sediment and Nutrient Removal at BSR..22
        Habitat and Biological Responses at BSR 23
        2019 Follow Up Report...23
    Current Study at Robinson Run and Ryerson State Park – Study 24
        Predictions 25
Chapter 2: Methodology 28
    Site Descriptions...28
    Water Chemistry...31
    Algal (Periphyton) Biomass 32
    Benthic Macroinvertebrates 34
    Fish Sampling...36
    Statistical Analyses 38

Chapter 3: Results..40

    Water Chemistry..40

    AFDM..44

    Chlorophyll...46

    Macroinvertebrates...50

    Fish..67

    SEM Model...82

Chapter 4: Discussion...86

    Water Chemistry..86

    AFDM..88

    Chlorophyll *a*...89

    Macroinvertebrates...91

    Fish..97

    SEM Model...99

Chapter 5: Conclusions...102

References...105

Appendix A: Illustrations and Maps of Sites.......................................115

Appendix B: Raw Biological Data..119

Chapter 1: Introduction

Stream Restoration Practices

Past anthropogenic activities can impair physical, chemical, and biological characteristics of streams, rivers, and lakes. Stream restoration projects aim to restore the natural structure and function of the stream to improve the water quality. Many stream restoration projects include efforts to stabilize banks, plant riparian buffers, create in stream habitat, and design natural streams with riffles, pools, and meanders to repair stream structure and function (Wohl and Merritts 2007). In several circumstances, these restoration techniques fail because they do not consider the stream as a dynamic system that changes overtime. Recently, floodplain reconnection with newly constructed or existing floodplains has emerged as a method of stream restoration, specifically in large rivers (Paillex et al. 2009, Stoffels et al.2014). Floodplain reconnection projects aim to restore hydrologic connectivity between the river and floodplain to promote sediment and nutrient retention (Roley et al. 2012). Many reconnection projects create wetland floodplains that retain nitrogen (N) and phosphorus (P) and promote nutrient cycling in soils more effectively than streams (Wolf et al 2015). The herbaceous vegetation in wetland floodplains assists in slowing water velocity and increasing water storage through the development of hydric soils (Noe and McMillan 2017) as well as trapping sediment and organic matter (Tiner 1987). In large rivers, floodplain reconnection projects create new habitat for fish and macroinvertebrates and can promote biological diversity (Stoffels et al. 2014). In rivers, restoring floodplains and creating an anabranching channels along the floodplain can maintain water and sediment flux without changing the channel gradient (Nanson and Knighton 1996). This restoration technique allows the stream channel to behave dynamically and change naturally overtime. Positive restoration response to floodplain reconnection can greatly improve water quality of impaired rivers, but responses can vary in smaller streams.

Controversy Surrounding Reconnection

In streams, particularly smaller first through third order streams, the restoration responses of floodplain reconnection can be unpredictable and controversial (Jones et al. 2015). Many argue that this project may not be the best management practice to promote nutrient retention due nutrient enrichment events following reconnection During European colonization in the 18th century, excessive logging and agricultural activities resulted in sediment eroding from upland hill slopes, burying pre-settlement streams, floodplains, and wetlands (Walter et al. 2007). Current and historical agricultural events, such as cattle and horse grazing, selective logging of valley side slopes, and resource extraction resulted in streams with high legacy sediment (Salvador et al. 1993, Figure 1). Legacy sediment built up over the years resulted in detached floodplains, bank incisions, and poor channel structure (Figure 1). Over the past decade, the Pennsylvania Department of Environmental Protection (PA DEP) has implemented floodplain reconnection methods to restore floodplains to their natural state. Floodplain reconnection has become a more widely accepted use of stream restoration, particularly in larger rivers (Wohl and Merritts 2007) but has been noted as controversial in smaller streams. In smaller order streams, legacy sediment from past activities may become a source of nutrients following floodplain reconnection (Jones et al. 2015). Grading the floodplain and filling catchments with legacy sediment can expose streams to nutrients (Walter et al. 2007) and can lead to excessive nutrient enrichment following reconnection. Increased nutrient enrichment following floodplain reconnection can impact biomass of periphyton, macroinvertebrates and fish taxa in unpredictable ways (Qu et al. 2019). Since wetland-stream complexes are rare, there is very little research present on the responses to biological communities in reconnection events. Several restoration projects are being implemented in western Pennsylvania, including at our study sites, Robinson Run and Ryerson Station State Park.

Current Floodplain Reconnection Projects

Current floodplain reconnection, or elevation, projects conducted by Pennsylvania Department of Environmental Protection (DEP) focus on a smaller category of streams (1st-3rd order). The projects conducted in Ryerson Station State Park and Robinson Run (western Pennsylvania) focus on reconnecting the stream to the pre-settlement floodplain elevation to create a wetland-stream complex. The state of PA has designated the area as a mitigation bank to restore these streams to pre-settlement conditions. These wetland-stream complexes existed in the pre-settlement era before excessive logging and agricultural activity buried them in legacy sediment (Walter et al. 2007). In two already completed projects (Robinson Run and Big Spring Run), PA DEP removed legacy sediment from the sites, graded the floodplain, and pushed sediment up into smaller catchments to level the floodplain (PA DEP Final Report 2013). Floodplain reconnection will be conducted in the same manner at Ryerson Station State Park. The reconnection projects plan to convert the high-gradient, nutrient poor streams into low-gradient streams with floodplain wetlands exhibiting increased sediment, nutrient, and water retention (First Pennsylvania Resource) that promotes nutrient cycling and improved water quality, equating to a healthier ecosystem. The end result is a habitat characterized by braided, anastomosing channel that flows through wetland floodplains as present in pre-settlement eras (Richardson et al. 2010). Such, a wetland-stream complex is expected to be stable for more than 200 years (Figure 2). Wetland-stream complexes are rare due to land use change from current and historic anthropogenic activities (Salvador et al1993), but there are some natural; for example, Great Marsh located in eastern Pennsylvania near Philadelphia. In the next few paragraphs, I will review some of the literature on biological trends that have resulted from floodplain reconnection events and nutrient enrichment in streams and wetlands.

Literature Review – Biological Responses to Floodplain Reconnection and Nutrient Enrichment

Periphyton

Periphyton, or algal communities surrounded by non-photosynthetic biofilms, use nutrients such as nitrate (N) and phosphate (P) to accumulate biomass. Floodplains tend to be highly productive (Olde et al. 2006) and can be a key depositional site for nutrients, such as carbon, nitrogen, and phosphorous (Tockner et al. 1999). Past studies of floodplain reconnection in smaller streams suggests that the process can lead to an increase in nutrients following reconnection (Jones et al. 2015) and subsequent periphyton proliferation. When determining nutrient enrichment events, many studies include two measures of algal biomass: ash-free dry mass and chlorophyll *a* levels.

Chlorophyll *a* levels can increase to 23 mg*m^{-2} and ash-free dry mass can increase to 900 mg*m^{-2} in nutrient enrichment events (Greenwood and Rosemond 2005). ADFM and chlorophyll *a* have also been found to relate positively with nutrient enrichment (Atazadeh et al. 2009), but in some situations, AFDM tends to not be as strongly correlating with nutrient levels. Nutrient levels and uptake tend to be higher in recently restored streams, correlating with chlorophyll *a* concentration (Newcomer et al. 2016). Typically, at sites that are undergoing an increase in nutrient levels, chlorophyll *a*, tends to increase (Nelson et al. 2013, Sharifi and Ghafori 2005). This may be a response to an exposure to nutrient enriched soils during restoration (Big Spring Run Research Results 2019). Algal biomass may increase in response to an increase in nitrogen and phosphorus but may vary depending on the ratio of nutrients and which is limiting production. In vernal pools, artificial nutrient enrichment increased phosphorus and chlorophyll *a* levels, but not nitrogen (Kido and Kneitel 2021). In agricultural streams, chlorophyll *a* increased with N and P enrichment, but AFDM did not vary with P enrichment (Taylor et al. 2020). In wetlands, nutrient enrichment tends to increase periphyton biomass also, but there have been instances where phosphorus enrichment has depleted periphyton biomass (Liston et al. 2008). In this case, calcareous periphyton mats were common (Florida Everglades) and decrease in biomass as filamentous cyanobacteria replaced them.

Part of the complexity of periphyton response to nutrient enrichment is related to other factors that also influence biomass accumulation. Light availability influences periphyton growth, as increased light availability can result in increased nutrient uptake and levels (Newcomer et al. 2016). Post-restoration streams tend to have open canopies, since much of the canopy area surrounding the stream is removed during restoration

construction. Open canopy streams also tend to have higher chlorophyll *a* and AFDM

levels (Fuller et al. 2008). Temperature and dissolved oxygen (Qu et al. 2019) are also

important, and standing algal biomass is also regulated by feeding of herbivorous

macroinvertebrates and fish (Fuller et al. 2008). Increases in periphyton growth levels can

improve nutrient retention, which is the goal of the floodplain reconnection project, but

nutrient enrichment can also lead to negative responses in macroinvertebrates and fish.

Macroinvertebrates

The response of macroinvertebrates to floodplain reconnection and nutrient

enrichment events are varied and depend on specifics of the region, site, and taxonomic

group of interest. Disturbance following stream restoration events can negatively impact

macroinvertebrates, particularly sensitive taxa. In some reconnected stream floodplain

systems on a main river channel in France, EPT taxa richness, or sensitive taxa, increased

(Paillex et al. 2007). In other reconnected floodplain, EPT taxa and MAIS scores

decreased immediately following reconnection, possibly due to disturbance of the

channel (Big Spring Run Research Results 2019, Smith et al. 2020). Disturbed sites may

result in a decrease in macroinvertebrate diversity, specifically sites with high sediment

phosphorous levels (Chipps et al. 2006). Feeding groups also show differing responses to

reconnection. Feeding groups such as collector filterers (CF) increased, but predators

decreased (Paillex et al. 2007). Chironomidae larvae, a collector-gatherer (CG), have

been found to increase and benefit from disturbance events and nutrient enrichment

(Fuller et al. 2008). Following disturbance events, changes to water chemistry or habitat can influence the community structure of macroinvertebrates. These results may vary due to the ability of certain taxa to reproduce and disperse quickly after reconnection events.

Nutrient enrichment can also cause varying macroinvertebrate responses. In streams, a common response is an increase in biomass of macroinvertebrates in response to high periphyton biomass (Justus et al. 2010), especially for grazers such as Gastropoda (Hill et al. 1992). Sraper-grazers (SC) can utilize periphyton biomass as a food source, and it is common to see an increase in scraper-grazers with an increase in periphyton biomass (Gjerløv and Richardson 2010). In wetlands, feeding groups such as collector/gatherers and grazers increased in biomass (Rader and Richardson 1994) as well as overall invertebrate density (Liston et al. 2008). Other feeding groups that utilize detritus material resulting from periphyton breakdown, such as collector/gatherers and filter feeders, can also increase in streams following enrichment events (Biggs et al. 2000). Invertebrate biomass may also be influenced by top-down regulation of fish.

Fish

Fish, usually the last taxa to return after a reconnection event, also can reflect varying responses to floodplain reconnection and resulting nutrient enrichment. In streams, biomass of fish detritivore/herbivore feeding groups are commonly expected to increase during levels of high periphyton biomass (Justus et al. 2010). However, nutrient enrichment can be detrimental to fish; densities of the herbivorous central stoneroller minnow (*Campostoma anomalum*) and the insectivorous orangethroat darter (*Etheostoma spectabile*) decreased in low level nutrient enrichment scenarios in Texas streams (Taylor 2011). Families such as Cyprinidae, many of which are herbivorous/omnivorous, show negative correlations with high total nitrogen and total phosphorus levels (Qu et al. 2019). Many fish, such as sand shiners (*Notropis stramenius*), bluegill (*Lepomis macrochirus*), fathead minnows (*Pimephales promelas*), green sunfish (*Leopmis cyanellus)* and yellow bullheads (*Ameiurus natalis*) are known to be highly tolerant and can be indicators of nutrient enrichment, while others can be indicators of low nutrients, including creek chubs (*Semotilus atromaculatus*), and blacknose dace (*Rhinichthys atratulus*) (Frey et al. 2011).

Fish assemblages in wetlands exhibit a similar response to nutrient enrichment. Smaller fish tend to increase abundance in highly enriched areas (Rader and Richardson 1994), particularly herbivorous and omnivorous fish (Sargeant et al. 2011). Fish in feeding groups such as herbivores (central stonerollers), insectivores, or omnivores (sand shiners) can be influenced by changes in either periphyton or macroinvertebrate biomass.

Trophic Level Interactions

Interactions between trophic levels of periphyton, macroinvertebrates, and fish occur in nearly every aquatic ecosystem. Many aquatic communities are structured by both 'top-down', where the upper-level of the food chain (fish/macroinvertebrates) controls primary production, or by 'bottom-up', where abiotic factors such as light or nutrients control primary production (Liboriussen et al. 2004). In nutrient enriched streams, autotrophic (algal) production is high and herbivorous (grazers, scrapers) macroinvertebrates that feed on periphyton dominate (Biggs et al. 2000). High correlations occur between Chironomids (collector/gatherers), grazer densities, and chlorophyll *a* levels (Biggs et al. 2000), but these interactions can change when fish are present. Liston et al. (2008) reported increased macroinvertebrate densities in response to phosphorus enrichment until periphyton mats were lost, then they decreased as a result of habitat loss and refuge from fish predation. Grazer macroinvertebrate densities decreased in the presence of insectivorous fish, therefore resulting in higher periphyton biomass due to a reduction in grazing (Biggs et al 2000). Thus, predicting the responses of periphyton, macroinvertebrates, and fish can be complex.

Abiotic Factors

Other environmental factors may impact biological responses. Habitat, temperature (Higgins et al. 2009), and drainage area (Foster and Lewis) influence how fish and macroinvertebrates respond to nutrient enrichment. A wide range of habitat can support diverse assemblages of macroinvertebrates and fish and promote interactions between trophic levels. There may be different responses in biological assemblages in nutrient enriched streams with poor habitat. Canopy cover can also heavily influence periphyton biomass, as it has been found that open canopy streams have a higher chlorophyll *a* and AFDM levels (Fuller et al. 2008).

Additionally, the nature of biotic and abiotic processes varies with stream size in a fairly predictable manner. This natural 'scaling' of stream systems was first recognized in the 1980's and described as the Stream/River Continuum Concept by Vannote and colleagues.

River Continuum Concept

In this study, stream size may heavily influence biological responses. The River Continuum Concept (RCC) describes a connected continuum between small, forested headwater streams to large, open rivers (Vannote et al. 1980). The continuum considers the changes in the abiotic and riparian characteristics along the continuum that result in a natural and somewhat natural gradient in biological communities and how resources flow in and out of the aquatic ecosystems. Small, headwater streams (order 1-3) are typically characterized by high gradients and are heavily shaded by dense forest canopy in eastern temperate forests that supplies the stream with leaf litter as the main carbon source. Photosynthetic rates and periphyton production are low. The macroinvertebrate community is typically dominated by shredders that utilize the course particulate organic matter (CPOM), and fish communities consist of insectivorous fish. As the stream width increases and gradients decrease (order 4-6) the canopy opens and light becomes an important abiotic factor. Photosynthetic rates increase and periphyton productivity is high. Leaf litter decreases, and the main carbon sources becomes downstream transport of fine particulate organic matter (FPOM). Macroinvertebrate communities are dominated by scraper-grazers that shear algae from surfaces and collectors that filter from organic transport or gather FPOM from sediments. The fish communities in larger order streams have higher densities, higher diversity and good representation of both insectivores and piscivores (Sheldon 1968). Any investigation into biological responses to stream restoration must be take into account the naturally occurring hydrological, chemical, and biological features along the stream continuum.

Big Spring Run – Case Study

A similar floodplain reconnection method was completed in Big Spring Run (BSR) in eastern Pennsylvania 8 to 9 years ago (PA DEP Final Report 2013). BSR is located in Lancaster County, PA in the Piedmont ecoregion. The Pennsylvania DEP selected BSR to be the site of this new floodplain restoration practice. The BSR project aimed to restore the stream to pre-settlement floodplain structure and function by restoring natural valley morphology and physical characteristics. Many streams in the Piedmont region were dammed for milling purposes, and as dams were breached, an accumulation of legacy sediment and bank incision occurred (Walter et al. 2007). As soil and legacy sediment are exposed by erosion along incised banks, high levels of phosphorus and nitrogen is released. At Big Spring Run, construction occurred during September-October 2011 and included floodplain grading, removal of legacy sediment, planting of commercial wetland seed, and addition of woody debris to restore the floodplain (citation). Ongoing ecological and geomorphic monitoring has been conducted at BSR, including researching pre- and post-restoration macroinvertebrate communities, habitat assessments, sediment source and nutrient flux assessment.

Sediment and Nutrient Removal at BSR

A main goal of the project at BSR was to remove the buildup of legacy sediment. 21,704 yd^3 of legacy sediment was removed to restore the floodplain to pre-settlement ground surface elevation. Approximately 63,669 lbs of total nitrogen and 50,498 lbs of total phosphorus were removed during restoration, with a goal of reducing in stream nutrient loads (PA DEP Final Report 2013). Previous strategies for restoration have focused on slowing or stopping stream bank erosion instead of restoring them to their resettlement state of meandering along a wetland valley bottom (Wohls and Merritts 2007). The BSR Report following post-restoration in the year 2012 concluded that sediment loading, erosion, and fine sediment deposition were significantly lower than pre-restoration, but no data on the changes of nutrient fluxes and nutrient load reductions were presented.

Habitat and Biological Responses at BSR

Rapid habitat and visual stream assessments were conducted pre-restoration and 8 months post-restoration. The project met its restoration goals in improving sediment deposition and bank stability but experienced reductions in riparian vegetation. Macroinvertebrate densities and diversity tended to decline over 3 years post restoration, with a loss of EPT taxa, or taxa sensitive to water pollutants. Gammaridae, Chironomidae, Simuliidae, Baetidae, Hydrospychidae, and Elmidae were abundant in both pre- and post-restoration sites. MAIS scores ranged from fair to poor in all post-restoration sites. Chironomidae and Baetidae were in abundance in post-restoration streams, which can indicate streams impacted by suspended sediment, possibly from construction disturbance. Sediment removal projects can tend to result in a decline in macroinvertebrates immediately following reconstruction (Gilkinson et al. 2005). Macroinvertebrate communities are expected to recover around 4 years post-restoration as the emergence of submerged aquatic vegetation will create habitat and cover for macroinvertebrate assemblages (PA DEP Final Report 2013).

2019 Follow Up Report

Recently, the group involved in restoring BSR released another report stating that small, anastomosing channels had developed 3-4 years post restoration and terrestrial laser surveys show very little erosion occurring (Big Spring Run Research Results 2019).

From pre- to post-restoration, suspended sediment concentrations decreased by 87%, suspended sediment load decreased by 71% and total phosphorus concentrations decreased by 79%. The report claims that the removal of legacy sediment and aquatic instream ecosystem restoration was 10.5 times more effective than implementing other best management practices (BMP's). Biological surveys were apparently also conducted, but very little data was providing regarding the changes pre- and post-restoration. The report did indicate an increase in diatom biodiversity, and an increase in diatom species associated with nutrient enrichment and more associated with low nutrients. The presence of the rosyside dace (*Clinosomus funduloides*) is also an indicator of improved water quality as it prefers colder headwater streams with rocky substrate and gravel riffles (Big Spring Run Research Results 2019). There was no report of the macroinvertebrate response in this document.

Current Study at Robinson Run and Ryerson State Park – Study

PA DEP has implemented several floodplain reconnection projects, including the project that was completed at Robinson Run and will be implemented at Ryerson Station State Park. PA DEP recruited the Voinovich School faculty and staff to investigate changes in hydrology, nutrient flux, flow, and water chemistry at Robinson Run and Ryerson State Park. This study focuses on biological responses at the study sites and how the stream continuum plays a role on the communities.

To address the goal of characterizing the long-term development of biological communities after restoration, we proposed to sample between 6 and 10 stream reaches representing pre-restoration (Ryerson Station State Park) and 3-4 years post-restoration

(Robinson Run). Study sites included two sizes of streams at each site: headwater sites (drainage areas < 3 mi^2) and wadeable streams (drainage areas 14-25 mi^2). Our original number and identity of study reaches was modified later due in part to covid restrictions on travel and other difficulties in the field. See Methodology for more detail (Table 1).

In this study, we measured nutrient (total N, total P) levels and correlated the periphyton biomass, macroinvertebrate, and fish biomass, and feeding group responses. Several abiotic factors were measured in this study, including conductivity, pH, temperatures, and canopy type. Very little research has been done correlating periphyton, macroinvertebrate, and fish assemblage responses to nutrient enrichment, specifically in wetland-stream complexes, so this research will provide insight into the biological responses to this new floodplain restoration method.

Predictions

The pre-restoration sites were expected to follow the RCC, as the smaller headwater sites (Kent Run, McNay Run, Poland Run) are high gradient, shaded, and have a dense forest canopy. These 2nd order streams flow into N. Dunkard Fork (3rd-4th order), which has little canopy cover and high light availability. In the post-restoration sites, the gradient from the headwater sites to the medium, wadeable reaches is predicted to deviate slightly from the RCC. The anastomosing (Molinari Trib, Molinari and Lebanik HW) and 2nd order sites (Beham) have been restored into an open, wetland-stream complex, consisting of a low gradient, no canopy cover but some overhanging vegetation, and higher light availability than is typical of the smaller order streams in the RCC. The larger sites (Molinari and Lebanik) are similar to N. Dunkard Fork, with an open canopy and not much overhanding vegetation, but may vary slightly in biological communities due to downstream transport of nutrients and organic matter.

Biological responses to nutrient enrichment, specifically in recently floodplain reconnected sites, can be unpredictable at times, as much of the literature shows. Higher recovery time is needed for restoration to repair ecosystem functions (Noe and McMillan 2017). Stream size also plays an important role in the responses of biological communities, as described by the River Continuum Concept (Vannote et al. 1980). At Ryerson Station State Park, the pre-restoration (control) sites, it is predicted that:

- Nutrient enrichment, chlorophyll a, and AFDM biomass will be very low in the smaller, 2nd order streams.

- Nutrient levels, chlorophyll a, and AFDM biomass will increase with the stream size (e.g. higher in N. Dunkard Fork).

- The biological communities will closely resemble those described by the RCC, with few collector-gatherers and scraper-grazers in the smaller orders, few herbivorous fish, and then increase in N. Dunkard Fork.

The post-restoration anastomosing, 2nd order, and wadeable streams biological communities and nutrient patterns are expected to deviate from the RCC due to the changes from a high-gradient, fast flowing stream to a slow-moving, low gradient wetland-stream complex, as well as potential exposure to legacy nutrients.

- The anastomosing and 2nd order streams are predicted to have higher chlorophyll a and AFDM levels due to increased light and decreased canopy cover.

- The wadeable sites are predicted to have the highest total nitrate and phosphate concentration, as well as chlorophyll *a* and AFDM biomass due to higher transport from upstream.

- Scraper-grazer and collector-gatherer abundance will be highest at post-restoration, wadeable sites as a response to the elevated nutrient gradient.

- Fish biomass and central stoneroller biomass will also be the highest at the wadeable, post-restoration sites due to an increase in macroinvertebrate biomass and algal biomass due to nutrient enrichment.

Chapter 2: Methodology

Site Descriptions

Ryerson Station State Park is located in Richhill Township, Greene County, PA (Figure 3) in the Western Alleghany Plateau ecoregion. The surrounding land is mixed between agricultural and forested lands. The agricultural lands are comprised of hay fields and pastureland. The forest cover consists of mixed hardwoods and successional forest with 76% forest cover, 5.74% urban development, 0.24% impervious area (U.S. Geological Survey 2012). The topography is very hilly, with steep-sloping valleys meet narrow hill tops with beds of sandstone, shale, and limestone. Soil types consist of non-hydric soils of silt loam, gravel, and clay. Mean basin elevation is 1299.2 feet, mean annual precipitation is 41 inches, and mean annual maximum temperature is 60°F. streams in Ryerson became degraded due to past anthropogenic activities, such as mining, poor agricultural practices, and logging. These practices resulted in excessive sedimentation and nutrient inputs, pre-settlement floodplains buried in legacy sediment, incised stream channels, and bank erosion (LandStudies, Inc.)

Robinson Run is located in southwestern Washington County, PA (Figure 4) in the Western Alleghany Plateau ecoregion. The surrounding land use consists of agricultural floodplain, successional forested habitat, and semi-degraded forest habitat. Forest covers 70% of the land, urban development covers 5.67%, and impervious surface covers 0.23% (U.S. Geological Survey 2012). The soils surrounding the watershed consist of hydric soils and non-hydric soils of silt loam. Water sources in Robinson consist of high groundwater tables and runoff due to steep topography. Mean basin elevation is 1245.7 feet, mean annual precipitation is 39 inches, and mean annual

maximum air temperature is 60°F (U.S. Geological Survey 2012). Many of the pre-

settlement floodplains of the streams in Robinson Run were buried in legacy sediment

due to current and past anthropogenic activity, such as road development, logging, and

cattle grazing. Particularly at Robinson Run, these activities resulted in stream

straightening and floodplain detachment, pushing streams against valley slides and

creating high gradient slopes. Before restoration, the streams suffered from bank

widening, excessive sedimentation and erosion rates, and poor aquatic habitat (First

Pennsylvania Resource).

Table 1

Catchments, Streams, and Correlating Sizes of Research Sites

Stage of restoration	Catchment	Size	Name of stream reach
Pre-restoration	Ryerson Station State Park	2nd order	Kent (2.66 mi^2)
		2nd order	Poland Run (0.5 mi^2)
		2nd order	McNay Run (0.0836 mi^2)
		Wadeable	North Dunkard Fork (24.3 mi^2)
3-4 years post-restoration	Robinson Run	Anastomosing	Molinari Tributary (0.83 mi^2)
		Wadeable	Molinari (14.2 mi^2)
		2nd order	Beham (3.02 mi^2)
		Wadeable	Lebanik (20.9 mi^2)
		Anastomosing	Molinari HW (<0.05 mi^2)
		Anastomosing	Lebanik HW (<0.05 mi^2)

Water Chemistry

Lab Chemistry - The Voinovich School faculty and Environmental Studies students conducted several sampling events over the past two years (July 2020, November 2020, March 2021, and May 2021) after training following PA DEP protocols. At the pre-restoration sites (Ryerson Station State Park) and the sties 3-4 years post-restoration sites (Robinson Run), samples were analyzed for total nitrogen (N) and phosphorus (P), dissolved N and P, metals, dissolved metals, and total organic carbon. Grab samples of water chemistry at mid-depth, mid-stream were collected at each site. Sample bottles (high density polyethylene HDPE) were rinsed 3 times with stream water prior to collecting samples. Stream water was filtered with a 0.45-micron filter for dissolved N and P, dissolved metals, and dissolved general chemistry. Nitric acid (HNO_3) was used to preserve the dissolved N and P and metals. The samples were be placed on ice and sent to the Department of Environmental Protection – Bureau of Laboratories in Harrisburg, PA. A Sontek FlowTracker® was used to measure discharge and velocity. Pore water, sediment samples from pit-traps, and salt tracing was also conducted during the sampling.

Field Chemistry - For this study, grab samples of water chemistry at mid-depth, mid-stream were collected at each at the time of biological sampling (July-Aug. 2020). All the sites pre- and post-restoration were measured for total N and P using a Hach field colorimeter (Model D900). Reactive orthophosphate was tested using PhosVer 3 reagents (Absorbic Acid method 8048, Hach Company 2013) and nitrate was tested using NitraVer 5 reagent (Cadmium Reduction method 8039, Hach Company 2013). Water samples were filtered prior to analysis and analyzed within 24 hours of collection. At Lebanik, we ran out of PhosVer 3 reagents, so phosphate was not measured. At Beham, nitrate readings read 0 mg/L on the day we were there. A handheld meter was used to measure field conductivity (μS/cm), pH, and temperature (°C) at each site during biological sampling. Canopy cover was assessed as either open or closed, depending on if the site was shaded by riparian vegetation and forest cover.

Algal (Periphyton) Biomass

Periphyton was sampled once during the summer throughout June to July 2020 for measurements of ash free dry mass (AFDM) and chlorophyll *a*. Chlorophyll *a* provides an isolated measurement of only the autotrophs present, while ash-free biomass includes both autotrophs and heterotrophs in the biofilm (and any associated organic matter). A modified version of the U.S. EPA Periphyton Monitoring Protocol for Multihabitat Sampling (Stevenson and Bahls 1999) was used to sample erosional sites (riffles) and depositional sites (stream edges) at each site (described below).

Erosional substrate was found at each stream in both pre- and post-restoration sites. To sample erosional substrates, 10 rocks were selected from riffles and a 7.56 cm^2 square cut from plastic lid was scrubbed with a toothbrush. The area was rinsed with a small amount of stream water using a turkey baster. The rinsate from the ten rocks was pooled into one sample, kept on ice until transported to the lab, and frozen at -20°C. For depositional substrates, a hollow plastic tube with a diameter of 5.5 cm was be pressed into sediment and a pancake spatula was slid underneath to extract the sediment to obtain a sample core of known area (approximately 8 cm depth). Ten sediment samples at three different depositional areas along the sampling reach were collected and held in gallon Ziploc bags, kept on ice until transported to the lab, and frozen. Due to time constraints and other considerations, only erosional substrate samples were measured and reported in this book.

The composite periphyton sample from each site was freeze dried (lyophilized) for 24-48 hours and weighed. Half of the material was used for chlorophyll a measurements and half for ash free dry mass (AFDM). For chlorophyll a, half the freeze-dried sample was later divided up into triplicate samples of 0.003 – 0.005 g. Samples were extracted in 10 ml of 90% acetone overnight for 24, then measured in a Turner TD-700 fluorometer (Turner Designs, Sunnyvale, CA) for chlorophyll a in Dr. Morgan Vis's laboratory at Ohio University following methods described in Drerup (2016). Samples were also corrected for pheophytin a using 0.1 M HCl. Several samples read over 300 µg/L and were diluted using a 10:1 ratio.

For AFDM, the remaining half of the lyophilized sample was weighed, then placed into pre-weighed porcelain crucibles. The crucible + original biomass was weighed, then combusted in a muffle furnace (450°C) for 4-5 hours to remove organic matter (Greenwood and Rosemond 2005). Samples were cooled, then reweighed and the mineral ash remaining was subtracted from the original mass to obtain the ash-free dry mass (AFDM) of the periphyton. Measurements for both chlorophyll a and AFDM were divided by half of the square area of the rock that was scrubbed to standardize the measurement per cm squared.

Benthic Macroinvertebrates

Macroinvertebrates were collected from natural substrate using a combination of kick and dip nets. A primary goal was to sample macroinvertebrates in the same array of microhabitats as periphyton for comparisons between periphyton biomass and functional feeding group (FFG) macroinvertebrate biomass and abundance. However, comparisons were only made between erosional periphyton and macroinvertebrates. In larger streams, both erosional (riffle) and depositional (stream edge) habitat, was sampled (Molinari, Molinari Trib, Lebanik, Beham, Kent, N. Dunkard Fork, Poland Run, McNay Run). A one-meter square kick net was used to sample three riffles (3 m^2 of habitat) and a 500-micron D-frame dip net was used to sample edge depositional habitat (20 dips or jabs). This protocol followed the rapid assessment methods developed for the MAIS (Macroinvertebrate Aggregate Index for Streams, Johnson 2007). In addition, depositional habitats were sampled using a Hess-style bucket sampler, a five-gallon plastic bucket with the bottom cut out following the protocol of Moore (2010). The bucket was embedded 1-2 inches into the substrate and a trowel was used to stir and suspend sediment and macroinvertebrates into the water column inside the bucket. A small aquarium net was dipped rapidly into the well-stirred sediment 20 times to capture suspended macroinvertebrate. Bucket samplers are well-suited to estimate the biomass of sediment-loving detritivores and microcrustaceans in slow waters of edges as well as wetlands. At Molinari HW and Lebanik HW, riffle habitat was absent, so only depositional substrate was sampled using Hess (bucket) samplers and 20 dips. The length of the sampling reach was the same as the fish sampling reach (see below).

Macroinvertebrate samples were composited into separate plastic containers for each sampling type. Macroinvertebrates were preserved in 70% ethanol for later inspection and identification. Total biomass (wet weights on electronic scale) was measured, and macroinvertebrate abundance, species richness, Shannon-Wiener diversity, and Pielou's Evenness were calculated using the vegan package (version 2.5-7) in R (Oksanen 2013) to compliment measures of periphyton biomass. Kick and dip samples were used to calculate % EPT and MAIS scores. % EPT (Ephemeroptera, Plecoptera, Trichoptera) was calculated to determine percentage of sensitive taxa at each site. MAIS scores were also calculated and range from 0 to 20 and are categorized as very poor (0 - 7), poor (8 - 11), good (12 - 15) or very good (16 - 18) conditions (Johnson 2007). To obtain wet weights, invertebrates were patted dry with a Kimwipe® for 1 minute prior to being weighed to the nearest ± 0.1 milligram (mg) (Wetzel et al 2005). The weights were added together for total biomass (mg) of all macroinvertebrates at each site.

A second measure of biomass of selected macroinvertebrates was also performed. Selected taxa to be measured include Heptageniidae mayflies (scraper-grazer), Hydropsychidae caddisflies (collector-filterer), Chironomidae midges (collector-gatherers), and Elmidae riffle beetles (scraper-grazers). Although not included in this study, a separate study was conducted to compare total dry biomass calculated from morphological measures and measured total wet biomass of the taxa mentioned above. Head capsule width of all macroinvertebrate's capsules and body lengths and widths were measured to the nearest 0.1 mm using a dissecting microscope (Miyasaka et al. 2008).

The head capsule width and body lengths were used in a regression model containing already published, taxa specific head width/body length to mass relationships to calculate the dry biomass of invertebrae (Stoffels et al. 2003). Use of both methods to calculate biomass provided two measures of biomass if one is inaccurate due to error.

Fish Sampling

A combination of electroshocking and seining (following a modified version of Ohio EPA's Volume III Standard Biological Field Sampling and Laboratory Methods for Assessing Fish and Macroinvertebrate Communities (Ohio EPA 2015)) was used to characterize composition of fish communities (richness, abundance, diversity, evenness, and biomass). The length of the fish sampling reach was dependent on the drainage area of the stream; a drainage area of less than 20 mi^2 requires a 100-meter sampling reach, while greater than 20 mi^2 requires a 150-meter sampling reach (Ohio EPA 2015). To scale these sites more consistently (14 mi^2-25mi^2), 100-meter reaches were used for all sites.

Sampling was conducted in five streams total: three larger wadeable streams, Molinari, Lebanik, and N. Dunkard Fork, and two 2nd order streams, Kent and Beham. In the larger streams, electroshocking was conducted. Amy Mackey (Voinovich School) led the electroshocking crew and provided a 150-meter longline, generator, and live well. At Kent, the stream reach was passed with a seine (4' x 10' with 1/4" mesh) 10 times. Seining was done in segments to systematically sample the entire area (Poos et al. 2007), so a pass was made every 10 meters for the 100-meter reach. The seine was placed with the weights in the water with two people on each end. The mesh net of the seine was faced towards a targeted habitat in the stream (riffle, stream edge, woody debris) and was shuffled towards that habitat as quickly as possible then lifted. At Beham, we used a modified version of dip netting for fish along the 100-meter reach because the width of the stream at the time was too wide to effectively use the seine. The fish were collected in 5-gallon buckets, identified, and weighed to the nearest gram if less than 1,000 grams, but measured to the nearest 25 grams if greater than 1,000 grams using a spring dial scale. Total biomass was calculated from the species biomass and was further divided into total biomass of each trophic designation (herbivore, omnivore, insectivore, carnivore, and generalist, Ohio E.P.A., 1987). Fish were identified and returned to the water unharmed.

The biomass of central stoneroller minnows (*Campostoma anomalum*) was examined as a possible indicator of nutrient enrichment. Central stonerollers solely graze on algae (herbivores) and their feeding can significantly reduce algal biomass. Stonerollers occupy benthic or pelagic waters in gravel substrate in pools and riffles and are tolerant to pollutants (Reisinger et al. 2011).

Fish sampling was performed according to an IACUC permit from Ohio University and collecting permits from the Pennsylvania Department of Wildlife.

Ohio EPA's Qualitative Habitat Evaluation Index (QHEI) was also used to record physical characteristics of stream and wetland habitats. These measures include visual assessments of channel form, woody debris, instream, substrate, and riparian vegetation. Amy Mackey (Voinovich School) conducted the habitat surveys at the sites that were sampled for fish.

Statistical Analyses

Drainage area and temperature were $\log_{10}$ transformed to meet normality assumptions prior to statistical analyses. Drainage area statistical analyses were run using square miles and later transformed to square kilometers on figures and graphs to stay consistent with metric measurements. Water pH, conductivity, total phosphate, total nitrate, chlorophyll *a* and AFDM were normally distributed. General linear models (linear correlations and regression analyses) were conducted to investigate possible associations between nutrient levels in water and characteristics of periphyton, macroinvertebrate and fish communities. Quadratic, cubic, and quartic polynomial models were used to determine the line of best fit for the data. Only linear, quadratic, and cubic models were used for the fish analyses due to a lack of data. Only linear and quadratic models were used for most of the AFDM regressions due to ineffective analyses. Regressions that were not significant were removed from figures. Initial linear models were developed with data from all sites, but sites were mainly stratified by drainage area to determine if fish and macroinvertebrate communities in the small streams are substantially different from those in the larger streams. ANOVA's and planned comparisons were used to examine differences of nutrient levels, periphyton biomass, macroinvertebrate and fish communities between stream size and pre- vs. post-restoration sites. Multiple stepwise regressions were used to regress water chemistry and stream characteristic variables against nutrient responses, and then nutrient responses against chlorophyll *a* and AFDM, and macroinvertebrate biomass.

. We conducted exploratory non-linear multidimensional ordinations (NMDS), PERMANOVA, and Bray Curtis similarity distances to describe the composition of macroinvertebrate and fish communities across the site categories using the vegan package in R. Finally, we explored the use of Structural Equation Modeling (SEM) to model relationships between environmental variables and periphyton/macroinvertebrate biomass using the lavaan package in R (Rosseel 2012).

Chapter 3: Results

Water Chemistry

Water chemistry data was variable across all sites, especially among the anastomosing and 2nd order streams. Temperatures ranged from 19.4 to 28.6 (°C), with the warmer temperatures coinciding with the open canopy, wadeable streams, and cooler temperatures coinciding with the smaller, shaded streams. Water pH values for all sites were neutral (7.5 to 8.6), and conductivity ranged from 322 to 637 μS/cm (Table 2). Nitrate and phosphate were also highly variable among all sites, ranging from 0 to 3.3 mg/L for total nitrate, and 0 to 2.75 mg/L for phosphate. Nitrate values of 3.3 mg/L and phosphate values of 2.75 mg/L represent the maximum value that can be determined by the Hach colorimeter. Kent Run, a shaded, 2nd order stream, had an unusually high nitrate level, while Beham, an open, 2nd order stream, had a high level of phosphate (Table 2).

Table 2

Water Chemistry for Stream Sites

Site	Temp (°C)	pH	Cond. (µS/cm)	Nitrate (mg/L)	Phosphate (mg/L)	Drainage Area (km^2)
Lebanik HW	27	7.4	497	0.4	0.48	0.13
Molinari HW	20.6	8.4	534	0.6	1.99	0.16
McNay Run	20.3	7.8	635	2.23	0.49	0.22
Poland Run	20.7	8.4	637	1.5	0.33	1.29
Molinari Trib	21.5	7.7	508	2.8	1.49	2.15
Kent Run	19.4	7.5	394	3.3	0.53	6.89
Beham	28.6	8.3	422	0*	2.75	7.82
Molinari	26	8.3	322	0.9	2.03	36.8
Lebanik	27.4	8.6	386	1.1	0*	54.1
N. Dunkard Fork	27.6	8.5	570	0.5	0.29	62.9

Note. Beham nitrate was non-detect, Lebanik phosphate was not measured.

Among pre- and post-restoration sites, two-way ANOVA's show that nitrate levels do not differ ($F_{1,8} = 1.7971$, $p = 0.22$), nor do phosphate levels ($F_{1,8} = 3.892$, $p = 0.08398$). A stepwise regression was run to determine which variables had the strongest effect on nitrate and phosphate. pH and conductivity were removed from the stepwise regressions due to small range. Out of a model including drainage area, temperature, phosphate, and canopy type, only temperature correlated significantly with nitrate levels ($R^2 = 0.574$, Adj. $R^2 = 0.521$, $F_{1,8} = 10.79$, $\beta = -0.758$, $p = 0.0011$). The final model output for phosphate included canopy cover but this variable did not significantly impact phosphate levels ($R^2 = 0.184$, Adj. $R^2 = 0.083$, $F_{1,8} = 1.81$, $\beta = 0.429$, $p = 0.215$). When drainage area alone was investigated, linear regressions show that neither nitrate ($F_{1,8} = 0.0469$, $p = 0.834$, Figure 1) nor phosphate ($F_{1,8} = 0.01095$, $p = 0.9192$, Figure 2) were impacted significantly.

Figure 1

Drainage Area and Total Nitrate (Linear: $R^2 = 0.0058$, $F_{1,8} = 0.047$, $p = 0.83$, Quadratic: $R^2 = 0.29$, $F_{2,7} = 1.46$, $p = 0.30$, Cubic: $R^2 = 0.34$, $F_{3,6} = 1.04$, $p = 0.44$, Quartic: $R^2 = 0.36$, $F_{4,5} = 0.69$, $p = 0.63$)

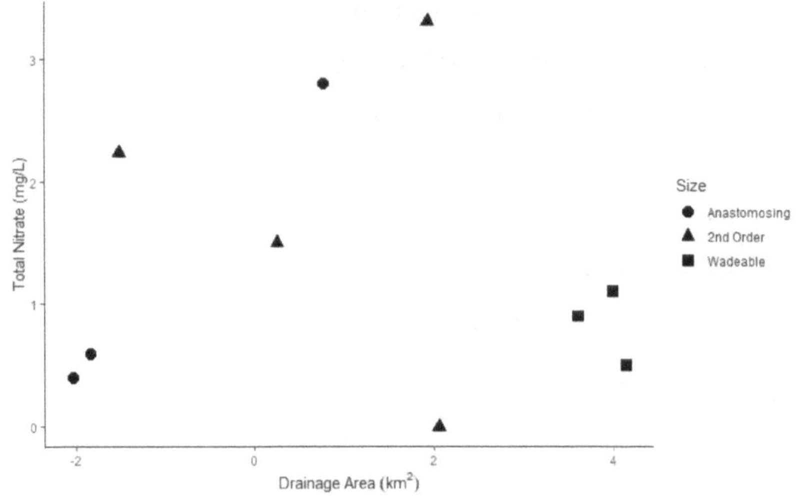

Figure 2

Drainage Area and Total Phosphate (Linear: $R^2 = 0.0014$, $F_{1,8} = 0.011$, $p = 0.92$, Quadratic: $R^2 = 0.080$, $F_{2,7} = 0.30$, $p = 0.75$, Cubic: $R^2 = 0.33$, $F_{3,6} = 1.01$, $p = 0.45$, Quartic: $R^2 = 0.39$, $F_{4,5} = 0.82$, $p = 0.57$).

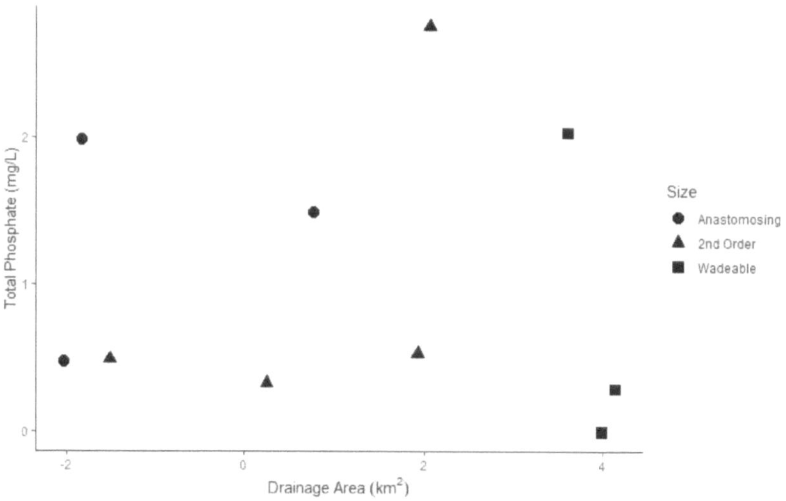

AFDM

Ash-free dry mass measured from erosional substrate ranged from 0.0023 to 0.02 g/cm². Molinari Trib. and McNay Run had the lowest AFDM. While AFDM differed between sites ($F_{9,20} = 3.33$, $p = 0.012$, Figure 3), there was no difference between pre- and post-restoration sites ($F_{1,28} = 0.026$, $p = 0.87$) based on two-way ANOVA's nor was there a correlation with drainage area ($F_{1,8} = 0.09$, $p = 0.77$, Figure 4) based on a linear regression. According to a stepwise regression including temperature, drainage area, nitrate, phosphate, canopy cover, and chlorophyll *a*, only nitrate was including in the final model output, but was not significant ($R^2 = 0.344$, Adj. $R^2 = 0.263$, $F_{1,8} = 4.2$, $\beta = -0.587$, $p = 0.074$).

Figure 3

AFDM Levels at the Study Sites, Categorized as Anastomosing, Second Order Headwater, and Larger Wadeable Streams (Boxplots represent Means and Standard Deviations of 3 Replicates of 10 Rocks).

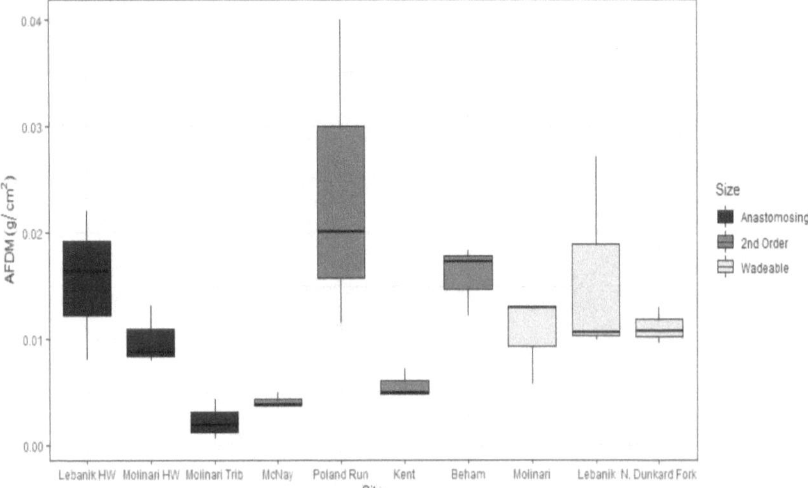

Figure 4

Drainage Area and Mean AFDM (Linear: R^2 = 0.011, $F_{1,8}$ = 0.09, p = 0.77, Quadratic: R^2 = 0.019, $F_{2,7}$ = 0.069, p = 0.93, Cubic: R^2 = 0.076, $F_{3,6}$ = 0.17, p = 0.92, Quartic: R^2 = 0.15, $F_{4,5}$ = 0.22, p = 0.91).

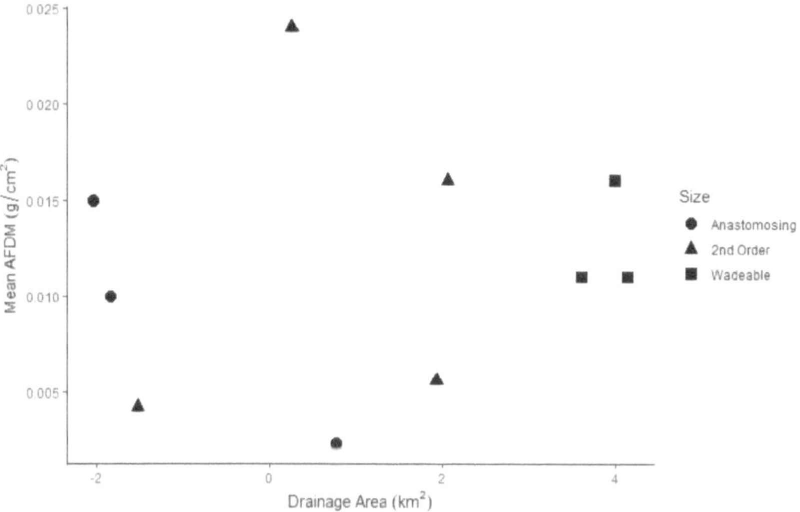

Chlorophyll

Chlorophyll *a* varied widely among the sites (3.49-200.2 µg). Molinari Trib exhibited the smallest biomass, while N. Dunkard Fork had the highest chlorophyll *a*. Chlorophyll *a* biomass varied among site ($F_{9,79}$ = 28.74, p < 0.0001, Figure 5) and drainage area ($F_{1,8}$ = 11.9, p = 0.0087, Figure 6). However, there was no difference in chlorophyll *a* levels between pre- and post-restoration sites ($F_{1,87}$ = 0.16, p = 0.69). According to a stepwise regression, several variables correlated with chlorophyll *a*, including drainage area (β = 0.78, p < 0.001), nitrate (β = -0.742, p < 0.001), and open canopy ($R^2 = 0.979$, Adj. $R^2 = 0.969$, $F_{1,8} = 93,44$, $\beta = -0.241$, $p = 0.023$). When analyzed with separate linear regressions, drainage area ($F_{1,8} = 11.9$, $p = 0.0087$, Figure 6) and nitrate ($F_{1,8} = 5.79$, $p = 0.043$, Figure 7) had a significant impact on chlorophyll *a,* but a two-way ANOVA showed canopy cover did not ($F_{1,8} = 1.82$, $p = 0.21$, Figure 8). Neither AFDM nor phosphate correlated with chlorophyll *a* response.

Figure 5

Chlorophyll a at the Study Sites, Categorized as Anastomosing, Second Order Headwater, and Larger Wadeable Streams (Boxplots Represent means and Standard Deviations of 3 Replicates of 10 Rocks).

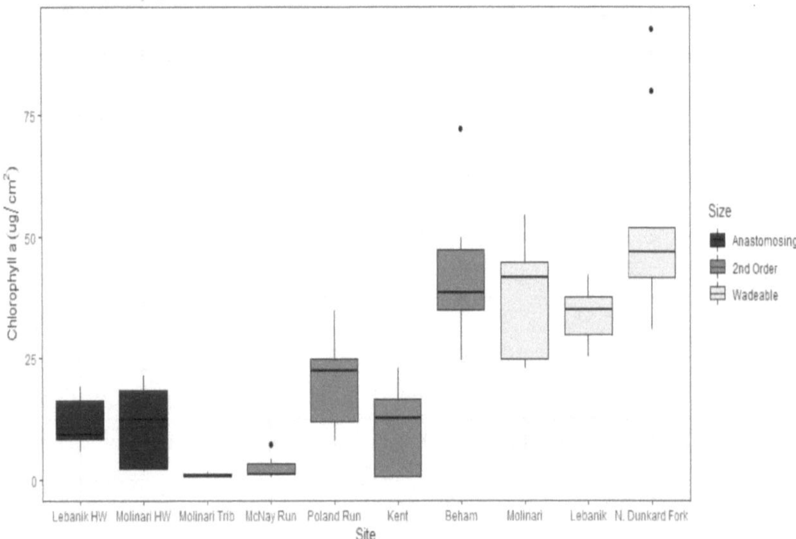

Figure 6

Drainage Area and Mean Chlorophyll a (Linear: $R^2 = 0.60$, $F_{1,8} = 11.9$, $p = 0.0087$, Quadratic: $R^2 = 0.68$, $F_{2,7} = 7.35$, $p = 0.019$, Cubic: $R^2 = 0.69$, $F_{3,6} = 4.35$, $p = 0.060$, Quartic: $R^2 = 0.69$, $F_{4,5} = 2.76$, $p = 0.15$)

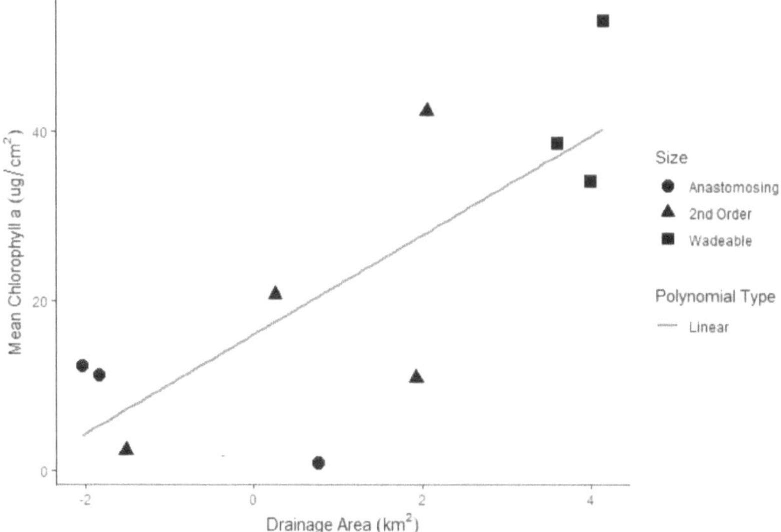

Figure 7

Total Nitrate and Mean Chlorophyll a (Linear: $R^2 = 0.42$, $F_{1,8} = 5.79$, $p = 0.043$, Quadratic: $R^2 = 0.43$, $F_{2,7} = 2.60$, $p = 0.14$, Cubic: $R^2 = 0.47$, $F_{3,6} = 1.74$, $p = 0.26$, Quartic: $R^2 = 0.53$, $F_{4,5} = 1.43$, $p = 0.35$).

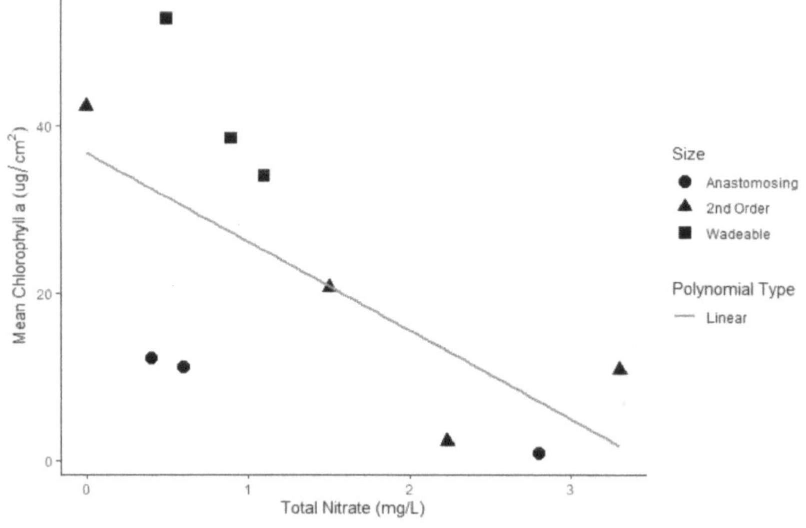

Figure 8

Mean Chlorophyll a between Closed and Open Canopy Cover ($F_{1,8}$ = 1.82, p = 0.214)

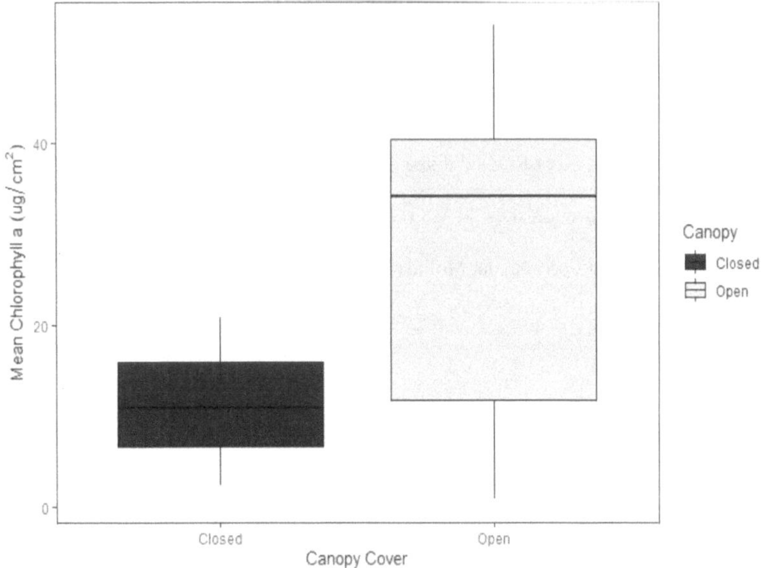

Macroinvertebrates

Macroinvertebrate community metrics varied more among streams rather than between pre- and post-restoration sites. Abundance from kick, dip, and bucket samples ranged between 111 and 536 individuals and increased with drainage area based on a linear regression ($F_{1,8}$ = 12.25, p = 0.0081) but a two-way ANOVA showed abundance did not differ between pre- and post-restoration sites ($F_{1,8}$ = 1.966, p = 0.1985). Richness ranged between 9 and 33 families and increased with an increase in drainage area ($F_{1,8}$ = 13.71, p = 0.006) but did not differ between pre- and post-restoration sites

($F_{1,8}$ = 0.1931, p = 0.672) according to a two-way ANOVA. Evenness ranged from 0.2319 to 0.3148, with the highest evenness at Molinari HW and the lowest at Kent Run. Diversity ranged from 1.1152 to 2.5989, with the highest diversity at Lebanik and the lowest diversity at Molinari Trib. Percentage of EPT, or sensitive taxa from kicks and dips, ranged from 0 to 74.5%. Lebanik HW had no EPT taxa present at the time of sampling, while Kent Run exhibited a high EPT percentage. Percent EPT taxa was lower at the post-restoration sites. MAIS scores ranged from 2 to 15, with Molinari Trib. having the lowest MAIS scored and Molinari, N. Dunkard Fork, and Kent Run all having a score of 15 (Table 3).

Table 3

Macroinvertebrate Taxa Biomass, Abundance, Richness, Evenness, Diversity, % EPT, and MAIS Scores among Stream Sites.

Site	Total Biomass (mg)	Total Abundance	Richness	Evenness	Diversity	% EPT	MAIS
Lebanik HW	1532	302	9	0.2801	1.2164	0	3
Molinari HW	2940	272	9	0.3148	1.3604	3.30882	5
McNay Run	1342.1	111	15	0.3017	2.0475	41.4414	7
Poland Run	1623.2	361	24	0.2347	1.9215	43.2133	14
Molinari Trib	3942.4	351	11	0.2485	1.1152	0.8547	2
Kent Run	5103.1	338	25	0.2319	1.9871	74.5562	15
Beham	11088.1	466	20	0.2407	1.6331	5.36481	9
Molinari	7628.8	536	29	0.262	2.5509	35.2612	15
Lebanik	11631.9	476	33	0.2515	2.5989	27.3109	13
N. Dunkard Fork	3422	379	20	0.2539	1.9006	49.8681	15

Total macroinvertebrate abundance and biomass increased with an increase in drainage area ($F_{1,8}$ = 12.25, p = 0.00081; $F_{1,8}$ = 7.009, p = 0.029, Figure 9). Macroinvertebrate abundance increased significantly with chlorophyll *a* ($F_{1,8}$ = 8.15, p = 0.021), while biomass also increased chlorophyll a ($F_{1,8}$ = 3.70, p = 0.091), although the response was not significant (Figure 11). Abundance and biomass showed no significant response to an increase in AFDM ($F_{1,8}$ = 1.82, p = 0.21; $F_{1,8}$ = 1.26, p = 0.28, Figure 10), but biomass appeared to slightly decrease as AFDM increased. According to a stepwise multiple regression, abundance correlates with drainage area (β = 0.79, p = 0.008) out of temperature, canopy cover, AFDM, and chlorophyll *a* (R^2 = 0.61, Adj. R^2 = 0.56, $F_{1,8}$ = 12.25). Drainage area also had the strongest effect and only significant effect on macroinvertebrate biomass response compared to chlorophyll, temperature, canopy cover, and AFDM (R^2 = 0.47, Adj. R^2 = 0.40, $F_{1,8}$ = 7.01, β = 0.68, p = 0.029).

Figure 9

a) Drainage Area and Total Macroinvertebrate Abundance (Linear: $R^2 = 0.61$, $F_{1,8} = 12.25$, $p = 0.00081$, Quadratic: $R^2 = 0.61$, $F_{2,7} = 2.32$, $p = 0.036$, Cubic: $R^2 = 0.70$, $F_{3,6} = 4.67$, $p = 0.052$, Quartic: $R^2 = 0.71$, $F_{4,5} = 3.031$, $p = 0.13$). b) Drainage Area (log transformed) and Total Macroinvertebrate Biomass (Linear: $R^2 = 0.47$, $F_{1,8} = 7.01$, $p = 0.029$, Quadratic: $R^2 = 0.47$, $F_{2,7} = 3.13$, $p = 0.11$, Cubic: $R^2 = 0.63$, $F_{3,6} = 3.45$, $p = 0.092$, Quartic: $R^2 = 0.65$, $F_{4,5} = 2.35$, $p = 0.19$).

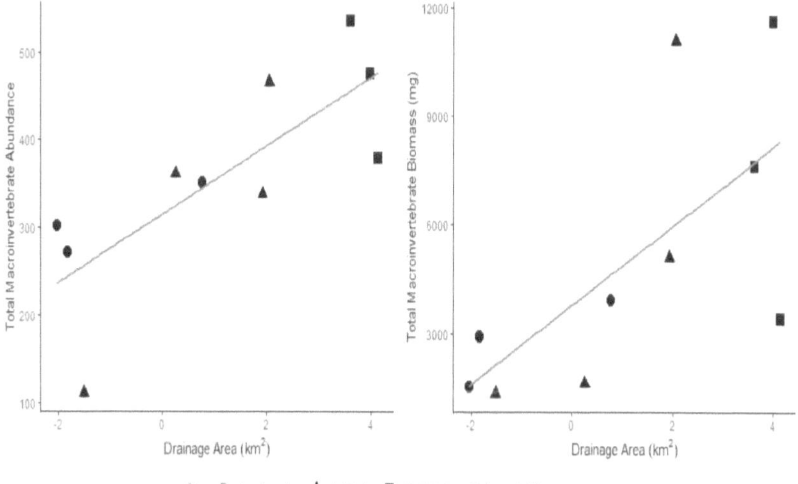

Figure 10

a) Mean AFDM and Total Macroinvertebrate Abundance (Linear: $R^2 = 0.19$, $F_{1,8} = 1.82$, $p = 0.21$, Quadratic: $R^2 = 0.23$, $F_{2,7} = 1.02$, $p = 0.41$, Cubic: $R^2 = 0.55$, $F_{3,6} = 2.39$, $p = 0.17$). b) Mean AFDM and Total Macroinvertebrate biomass (Linear: $R^2 = 0.14$, $F_{1,8} = 1.26$, $p = 0.28$, Quadratic: $R^2 = 0.14$, $F_{2,7} = 0.56$, $p = 0.60$, Cubic: $R^2 = 0.18$, $F_{3,6} = 0.43$, $p = 0.74$)

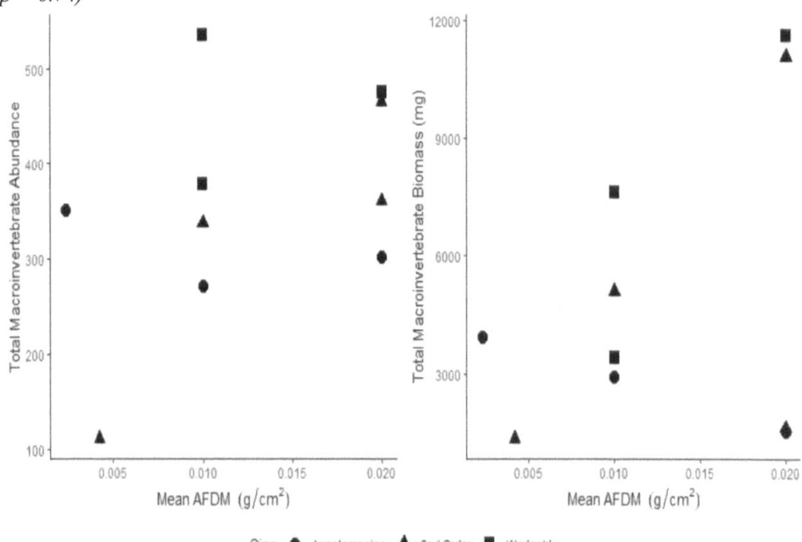

Figure 11

a) Mean chlorophyll a and total macroinvertebrate abundance (Linear: $R^2 = 0.50$, $F_{1,8} = 8.15$, $p = 0.021$, Quadratic: $R^2 = 0.63$, $F_{2,7} = 5.89$, $p = 0.32$, Cubic: $R^2 = 0.73$, $F_{3,6} = 5.44$, $p = 0.038$, Quartic: $R^2 = 0.73$, $F_{4,5} = 3.456$, $p = 0.1$). b) Mean chlorophyll a and total macroinvertebrate biomass (Linear: $R^2 = 0.32$, $F_{1,8} = 3.70$, $p = 0.091$, Quadratic: $R^2 = 0.40$, $F_{2,7} = 2.32$, $p = 0.17$, Cubic: $R^2 = 0.73$, $F_{3,6} = 5.45$, $p = 0.038$, Quartic: $R^2 = 0.77$, $F_{4,5} = 4.21$, $p = 0.074$).

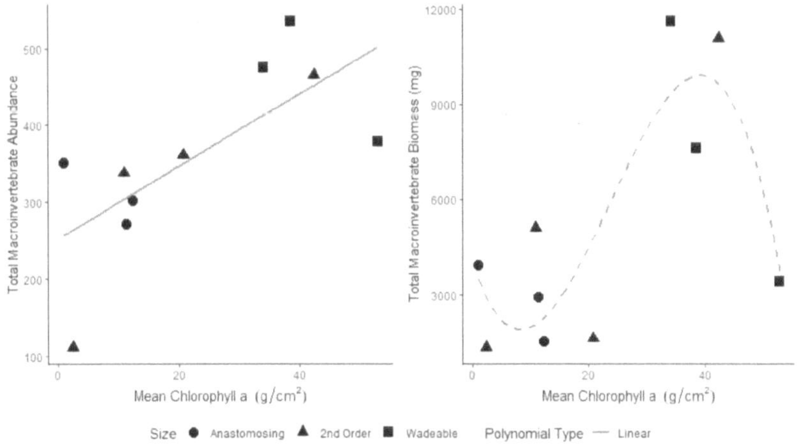

Macroinvertebrate communities were analyzed based on stream sites and how strongly these sites and communities correlate with pH, temperature, conductivity, nitrate, phosphate, canopy cover, drainage area, AFDM, and chlorophyll *a*. According to a permutational analysis (PERMANOVA) based on distance matrices, macroinvertebrates communities are different among stream sizes ($F_{2,7} = 3.03$, $p = 0.009$) and between pre-and post-restoration sites ($F_{1,8} = 2.80$, $p = 0.025$, Figure 13).

Figure 12

Macroinvertebrate NMDS Ordination (stress = 0.08027)

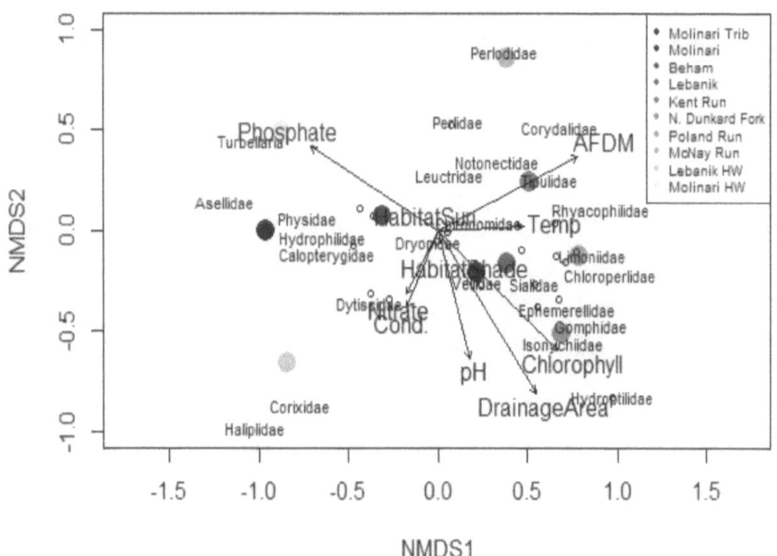

Bray-Curtis dissimilarities show that many of the sites exhibit dissimilar family composition (closer to one depicts a higher dissimilarity), but there are some sites that are similar in composition. Lebanik and Molinari display a Bray-Curtis value of 0.38, Molinari HW and Lebanik HW have a dissimilarity of 0.44, and Molinari HW and Molinari Trib are the most similar out of all the sites, with an index of 0.18. Poland Run and Lebanik HW are the most dissimilar sites, with an index of 0.99 (Figure 14). Physid snails were the only macroinvertebrate present at all of the sites. Molinari and Lebanik both exhibited a high abundance of Elmidae, Hydropsychidae, and Physidae families.

Molinari Trib., Molinari HW, and Lebanik HW, all anastomosing stream-wetland complexes, had a high abundance of Asellidae, Physidae, and Turbellaria. Beham, a 2nd order post-restoration site, also had a high number of Physidae and Turbellaria, as well as Elmidae riffle beetles. The pre-restoration, 2nd order sites exhibited a higher number of EPT taxa, including Perlidae, Chloroperlidae, Leuctridae, and Capniidae. N. Dunkard Fork had the highest abundance of Hydropsyhidae out of all of the sites, as well as high numbers of Elmidae and Chironomidae.

Figure 13

Macroinvertebrate Bray-Curtis Dissimilarities between Sites

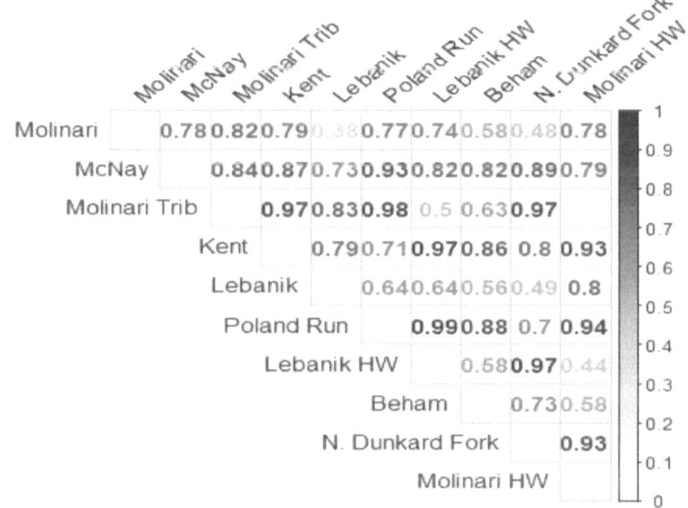

Total abundance and wet biomass of collector-filterer (CF), collector-gatherer (CG), and scraper-grazer (SC) feeding groups were analyzed to determine responses to changes in periphyton levels and drainage area. According to linear regressions, drainage area strongly correlated with CF abundance (Figure 14a) and biomass (Figure 15a), and CG biomass (Figure 17a). Chlorophyll *a* strongly correlated with CF abundance (Figure 14c) and biomass (Figure 15c), as well as SC abundance (Figure 18c). AFDM did not correlate significantly with any feeding group abundance or biomass (Table 4).

Table 4

Linear Regressions between CF/CG/SC Abundance and Biomass with Drainage Area, AFDM, and Chlorophyll a

Regression	R^2	$F_{1,8}$	P value
CF abundance ~ drainage area	0.55	9.69	0.014
CF abundance ~ AFDM	0.013	0.11	0.75
CF abundance ~ chlorophyll *a*	0.61	12.65	0.0074
CF biomass ~ drainage area	0.55	9.65	0.015
CF biomass ~ AFDM	0.0078	0.063	0.81
CF biomass ~ chlorophyll *a*	0.59	11.65	0.0092
CG abundance ~ drainage area	0.016	0.61	0.73
CG abundance ~ AFDM	0.02	0.17	0.69
CG abundance ~ chlorophyll *a*	0.02	0.16	0.070
CG biomass ~ drainage area	0.43	6.07	0.039
CG biomass ~ AFDM	0.029	0.24	0.64
CG biomass ~ chlorophyll *a*	0.34	1.67	0.078
SC abundance ~ drainage area	0.24	2.52	0.15
SC abundance ~ AFDM	0.093	0.82	0.39
SC abundance ~ chlorophyll *a*	0.45	6.56	0.033
SC biomass ~ drainage area	0.17	1.67	0.23
SC biomass ~ AFDM	0.00036	0.0029	0.96
SC biomass ~ chlorophyll *a*	0.13	1.16	0.31

Several stepwise regressions with CF/CG/SF abundance and biomass were ran including temperature, drainage area, canopy cover, AFDM, and chlorophyll *a*. Chlorophyll *a* elicited the strongest impact on CF abundance ($R^2 = 0.613$, Adj. $R^2 =$

0.564, $F_{1,8}$ = 12.65, β = 0.783, p = 0.007) and biomass (R^2 = 0.770, Adj. R^2 = 0.542, $F_{1,8}$ = 11.65, β = 0.77, p = 0.009). There were no variables that impacted CG abundance significantly, but open canopy was included in the model (R^2 = 0.188, Adj. R^2 = 0.087, $F_{1,8}$ = 1.86, β = 0.434, p = 0.21). Drainage area seemed to impact CG biomass significantly (R^2 = 0.431, Adj. R^2 = 0.360, $F_{1,8}$ = 6.065, β = 0.657, p = 0.039). Temperature had the strongest impact on SC abundance (R^2 = 0.687, Adj. R^2 = 0.722, $F_{1,8}$ = 20.77, β = 0.85, p = 0.002), while the best model output for SC biomass included drainage area, but was not significant (R^2 = 0.172, Adj. R^2 = 0.069, $F_{1,8}$ = 1.67, β = 0.415, p = 0.233).

Figure 14

a) Drainage Area and CF Abundance (Linear: $R^2 = 0.55$, $F_{1,8} = 9.69$, $p = 0.014$, Quadratic: $R^2 = 0.77$, $F_{2,7} = 11.57$, $p = 0.006^*$, Cubic: $R^2 = 0.85$, $F_{3,6} = 11.16$, $p = 0.007^*$, Quartic: $R^2 = 0.86$, $F_{4,5} = 7.87$, $p = 0.022$). b) Mean AFDM and CF Abundance (Linear: $R^2 = 0.013$, $F_{1,8} = 0.11$, $p = 0.75$, Quadratic: $R^2 = 0.16$, $F_{2,7} = 0.66$, $p = 0.54$, Cubic: $R^2 = 0.22$, $F_{3,6} = 0.57$, $p = 0.65$, Quartic: $R^2 = 0.34$, $F_{4,5} = 0.63$, $p = 0.66$). c) Mean Chlorophyll a and CF Abundance (Linear: $R^2 = 0.61$, $F_{1,8} = 12.65$, $p = 0.0074$, Quadratic: $R^2 = 0.73$, $F_{2,7} = 9.74$, $p = 0.0095$, Cubic: $R^2 = 0.78$, $F_{3,6} = 7.06$, $p = 0.021$, Quartic: $R^2 = 0.87$, $F_{4,5} = 8.46$, $p = 0.019$)

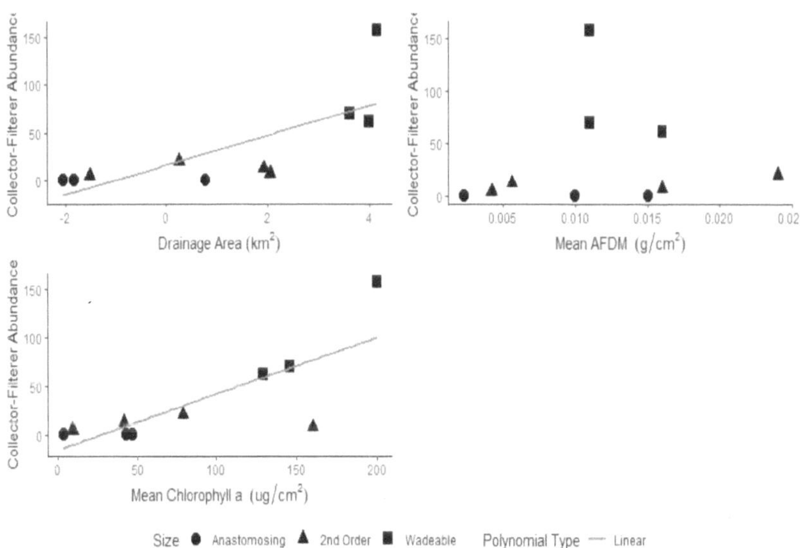

Figure 15

a) Drainage Area and CF Biomass (Linear: $R^2 = 0.55$, $F_{1,8} = 9.65$, $p = 0.015$, Quadratic: $R^2 = 0.77$, $F_{2,7} = 11.61$, $p = 0.0060^*$, Cubic: $R^2 = 0.84$, $F_{3,6} = 10.21$, $p = 0.009^*$, Quartic: $R^2 = 0.87$, $F_{4,5} = 8.65$, $p = 0.018$). b) Mean AFDM and CF Biomass (Linear: $R^2 = 0.0078$, $F_{1,8} = 0.063$, $p = 0.81$, Quadratic: $R^2 = 0.17$, $F_{2,7} = 0.74$, $p = 0.51$, Cubic: $R^2 = 0.23$, $F_{3,6} = 0.58$, $p = 0.65$, Quartic: $R^2 = 0.31$, $F_{4,5} = 0.56$, $p = 0.70$). c) Mean Chlorophyll a and CF Biomass (Linear: $R^2 = 0.59$, $F_{1,8} = 11.65$, $p = 0.0092$, Quadratic: $R^2 = 0.73$, $F_{2,7} = 9.44$, $p = 0.01$, Cubic: $R^2 = 0.80$, $F_{3,6} = 7.76$, $p = 0.017$, Quartic: $R^2 = 0.87$, $F_{4,5} = 8.48$, $p = 0.019$)

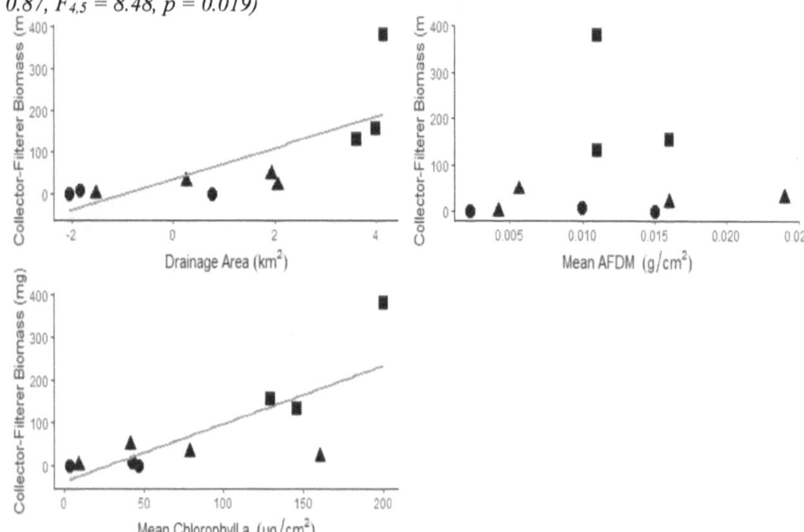

Figure 16

a) Drainage Area and CG Abundance (Linear: $R^2 = 0.016$, $F_{1,8} = 0.61$, $p = 0.73$, Quadratic: $R^2 = 0.16$, $F_{2,7} = 0.66$, $p = 0.55$, Cubic: $R^2 = 0.19$, $F_{3,6} = 0.48$, $p = 0.71$, Quartic: $R^2 = 0.30$, $F_{4,5} = 0.54$, $p = 0.71$). b) Mean AFDM and CG Abundance (Linear: $R^2 = 0.021$, $F_{1,8} = 0.17$, $p = 0.69$, Quadratic: $R^2 = 0.13$, $F_{2,7} = 0.53$, $p = 0.61$, Cubic: $R^2 = 0.30$, $F_{3,6} = 0.85$, $p = 0.52$, Quartic: $R^2 = 0.68$, $F_{4,5} = 2.68$, $p = 0.15$). c) Mean Chlorophyll a and CG Abundance (Linear: $R^2 = 0.020$, $F_{1,8} = 0.1632$, $p = 0.70$, Quadratic: $R^2 = 0.023$, $F_{2,7} = 0.084$, $p = 0.92$, Cubic: $R^2 = 0.090$, $F_{3,6} = 0.20$, $p = 0.89$, Quartic: $R^2 = 0.16$, $F_{4,5} = 0.24$, $p = 0.91$).

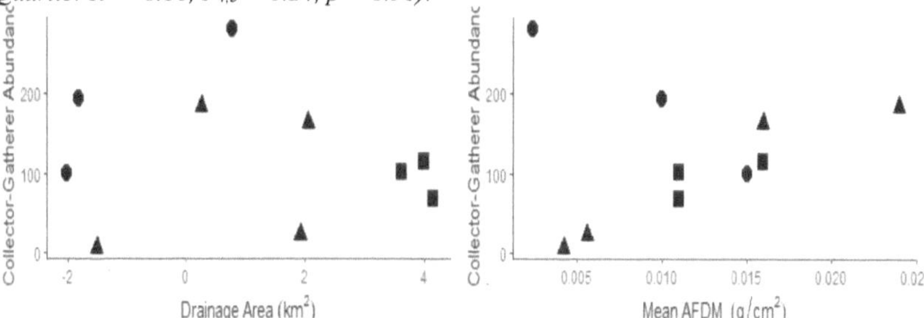

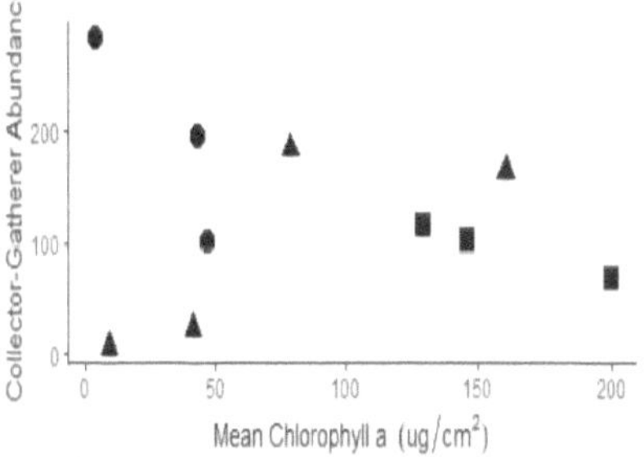

Figure 17

a) Drainage Area and CG Biomass (Linear: $R^2 = 0.43$, $F_{1,8} = 6.07$, $p = 0.039$, Quadratic: $R^2 = 0.70$, $F_{2,7} = 8.01$, $p = 0.016*$, Cubic: $R^2 = 0.79$, $F_{3,6} = 7.33$, $p = 0.020$, Quartic: $R^2 = 0.82$, $F_{4,5} = 5.79$, $p = 0.041$). b) Mean AFDM and CG Biomass (Linear: $R^2 = 0.029$, $F_{1,8} = 0.24$, $p = 0.64$, Quadratic: $R^2 = 0.17$, $F_{2,7} = 0.70$, $p = 0.53$, Cubic: $R^2 = 0.19$, $F_{3,6} = 0.45$, $p = 0.72$, Quartic: $R^2 = 0.18$, $F_{4,5} = 0.28$, $p = 0.88$). c) Mean Chlorophyll a and CG Biomass (Linear: $R^2 = 0.34$, $F_{1,8} = 1.67$, $p = 0.078$, Quadratic: $R^2 = 0.35$, $F_{2,7} = 1.89$, $p = 0.22$, Cubic: $R^2 = 0.35$, $F_{3,6} = 1.09$, $p = 0.42$, Quartic: $R^2 = 0.51$, $F_{4,5} = 1.31$, $p = 0.38$)

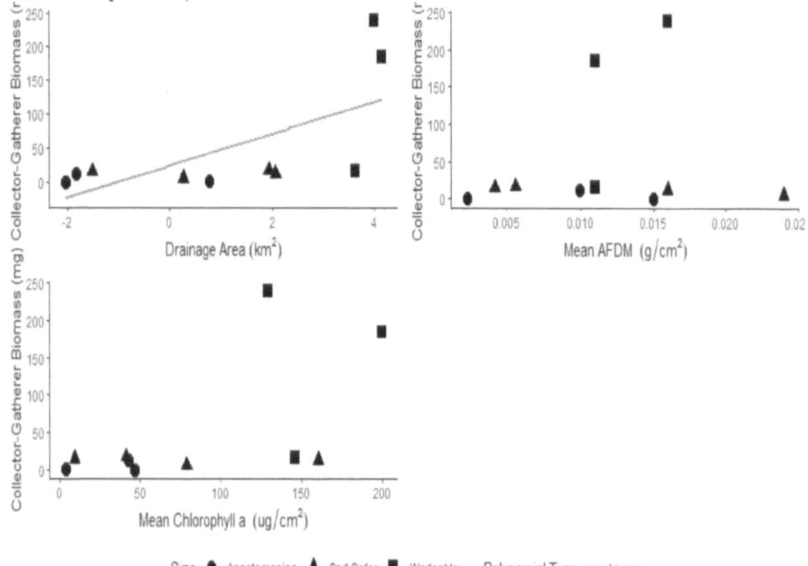

Figure 18

a) Drainage Area and SC Abundance (Linear: $R^2 = 0.24$, $F_{1,8} = 2.52$, $p = 0.15$, Quadratic: $R^2 = 0.30$, $F_{2,7} = 1.52$, $p = 0.28$, Cubic: $R^2 = 0.57$, $F_{3,6} = 2.62$, $p = 0.15$, Quartic: $R^2 = 0.59$, $F_{4,5} = 1.79$, $p = 0.27$). b) Mean AFDM and SC Abundance (Linear: $R^2 = 0.093$, $F_{1,8} = 0.82$, $p = 0.39$, Quadratic: $R^2 = 0.51$, $F_{2,7} = 3.65$, $p = 0.082$, Cubic: $R^2 = 0.66$, $F_{3,6} = 3.90$, $p = 0.074$, Quartic: $R^2 = 0.68$, $F_{4,5} = 2.67$, $p = 0.16$). c) Mean Chlorophyll a and SC Abundance (Linear: $R^2 = 0.45$, $F_{1,8} = 6.56$, $p = 0.033$, Quadratic: $R^2 = 0.51$, $F_{2,7} = 3.65$, $p = 0.088$, Cubic: $R^2 = 0.74$, $F_{3,6} = 5.63$, $p = 0.035$, Quartic: $R^2 = 0.84$, $F_{4,5} = 6.60$, $p = 0.031$).

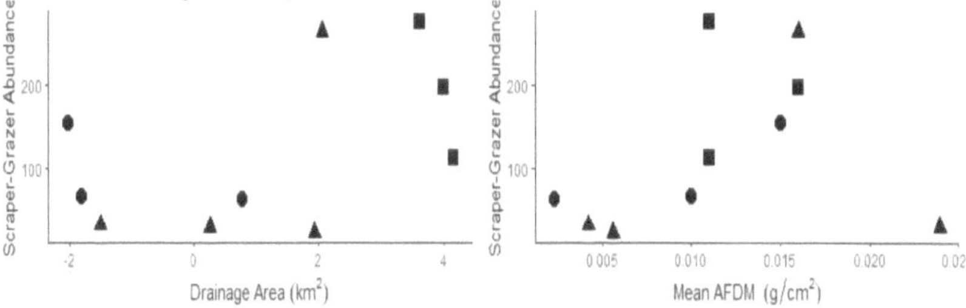

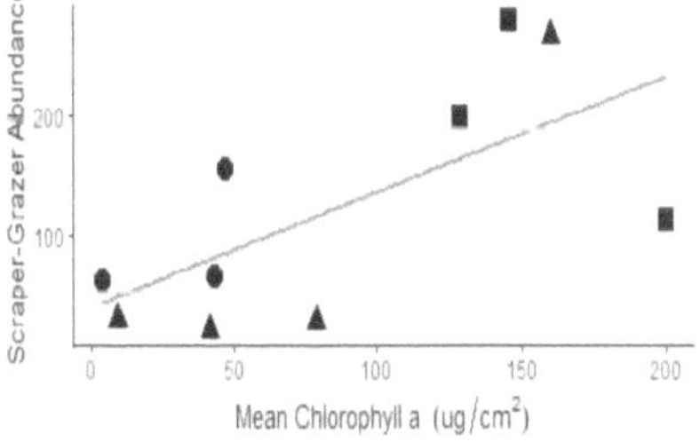

Figure 19

a) Drainage Area and SC Biomass (Linear: $R^2 = 0.17$, $F_{1,8} = 1.67$, $p = 0.23$, Quadratic: $R^2 = 0.19$, $F_{2,7} = 82$, $p = 0.48$, Cubic: $R^2 = 0.27$, $F_{3,6} = 0.73$, $p = 0.57$, Quartic: $R^2 = 0.56$, $F_{4,5} = 1.61$, $p = 0.30$). b) Mean AFDM and SC Biomass (Linear: $R^2 = 0.00036$, $F_{1,8} = 0.0029$, $p = 0.96$, Quadratic: $R^2 = 0.093$, $F_{2,7} = 0.36$, $p = 0.71$, Cubic: $R^2 = 0.16$, $F_{3,6} = 0.37$, $p = 0.78$, Quartic: $R^2 = 0.26$, $F_{4,5} = 0.45$, $p = 0.77$). c) Mean Chlorophyll a and SC Biomass (Linear: $R^2 = 0.13$, $F_{1,8} = 1.16$, $p = 0.31$, Quadratic: $R^2 = 0.17$, $F_{2,7} = 0.70$, $p = 0.53$, Cubic: $R^2 = 0.30$, $F_{3,6} = 0.86$, $p = 0.51$, Quartic: $R^2 = 0.32$, $F_{4,5} = 0.58$, $p = 0.69$).

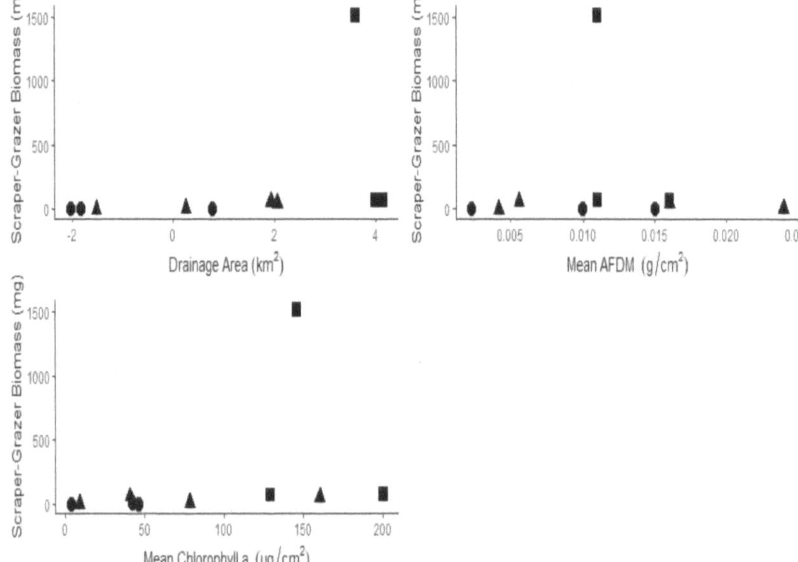

Fish

Fish sampling occurred at 5 out of the 10 sites; five streams were too small to support fish communities. Total biomass ranged from 162 to 13401 grams, with Beham having the lowest biomass and N. Dunkard Fork with the highest. Total abundance ranged from 162 to 3489, and richness ranged from 7 to 16, with Beham having the lowest abundance and richness, and N. Dunkard Fork having the highest. Evenness ranged between 0.27 to 0.32, and diversity ranged from 1.22 to 1.97. Beham exhibited the highest evenness and lowest diversity, Kent Run exhibited to lowest evenness, and N. Dunkard Fork had the highest diversity (Table 5).

Table 5

Fish Total Biomass, Abundance, Richness, Evenness, and Diversity at the Five Sites Large Enough to Support Fish

Site	Drainage Area (km^2)	Restoration	Total Biomass (g)	Total Abundance	Richness	Evenness	Diversity
Kent Run	6.89	Pre	240	166	13	0.27049	1.6532
Beham	7.82	Post	162	162	7	0.31652	1.2193
Molinari	36.8	Post	3553	1144	14	0.30199	1.9306
Lebanik	54.1	Post	2887	1536	13	0.30038	1.8694
N. Dunkard Fork	62.9	Pre	13401	3489	16	0.28475	1.9662

Bray-Curtis dissimilarity indices are low between Lebanik and Molinari, the two wadeable restored reaches, indicating these two sites are very similar in fish species composition. N. Dunkard Fork also has a similar species composition to Lebanik. N. Dunkard Fork and Kent exhibit the highest dissimilarity (Figure 20). Central stoneroller minnows, creek chubs, white suckers (*Catostomus commersonii*), bluntnose minnows (*Pimephales notatus*), johnny darters (*Etheostoma nigrum*), and fantail darters (*Etheostoma flabellare*) were present at all of the sites. Lebanik and Molinari both had a high abundance of stoneroller minnows, white suckers and fantail darters. Both also had stonecat madtoms, not found at any of the other sites.

N. Dunkard Fork had the highest abundance of central stonerollers, 1330, as well as a

high abundance of bluntnose minnows, rainbow darters (*Etheostoma caeruleum*), and

striped shiners (Table 10, Appendix B)

Figure 20

Fish Bray-Curtis Dissimilarities between Streams

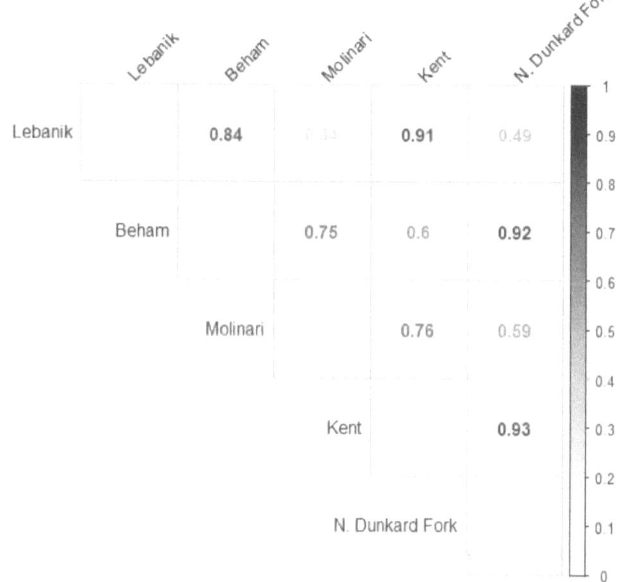

According to a PERMANOVA, fish species composition did not differ between

stream sizes ($F_{1,3}$ = 4.86, p = 0.1) or between pre- and post-restoration sites ($F_{1,3}$ = 0.64, p

= 0.6, Figure 13). The NMDS ordination was run in the same sense as the macroinvertebrate NMDS, including the same variables.

Figure 21

Fish NMDS Ordination

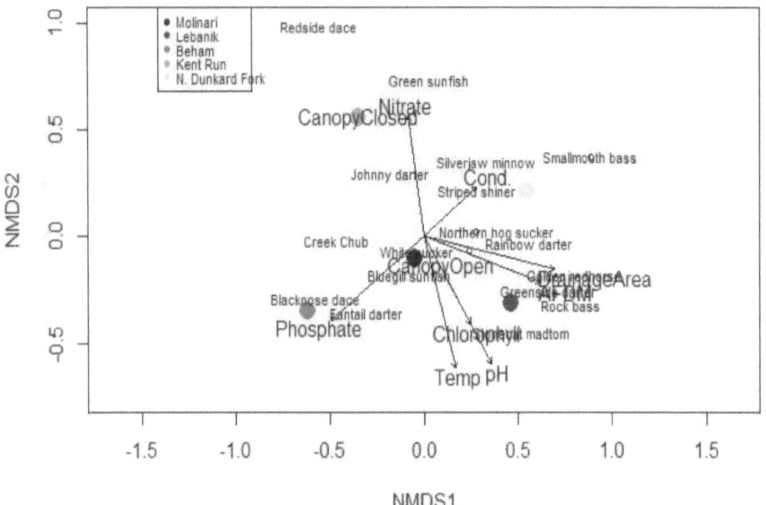

Linear regressions showed fish abundance and biomass both increased as drainage area increased, although not significantly ($F_{1,3} = 7.37$, p = 0.073; $F_{1,3} = 3.38$, p = 0.16, Figure 22). Chlorophyll *a* ($F_{1,3} = 2.57$, p = 0.21, Figure 24b) and AFDM ($F_{1,3} = 0.65$, p = 0.48, Figure 23b) also did not appear to have a significant impact on total fish biomass. Fish abundance did not increase with AFDM ($F_{1,3} = 0.30$, p = 0.62, Figure 23a), or with chlorophyll *a* ($F_{1,3}$ = 2.53, p = 0.21, Figure 24a). There was no significant relationship

with total macroinvertebrate biomass and total insectivore biomass ($F_{1,3}$ = 0.42, p = 0.56),

Figure 25).

Figure 22

a) Drainage Area and Total Fish Abundance (Linear: R^2 = 0.71, $F_{1,3}$ = 7.37, p = 0.073,
Quadratic: R^2 = 0.87, $F_{2,2}$ = 6.91, p = 0.073, Cubic: R^2 = 0.63, $F_{3,1}$ = 3.07, p = 0.39). a)
Drainage Area and Total Fish Biomass (Linear: R^2 = 0.53, $F_{1,3}$ = 3.38, p = 0.16,
Quadratic: R^2 = 0.71, $F_{2,2}$ = 2.42, p = 0.29, Cubic: R^2 = 0.76, $F_{3,1}$ = 1.06, p = 0.60)

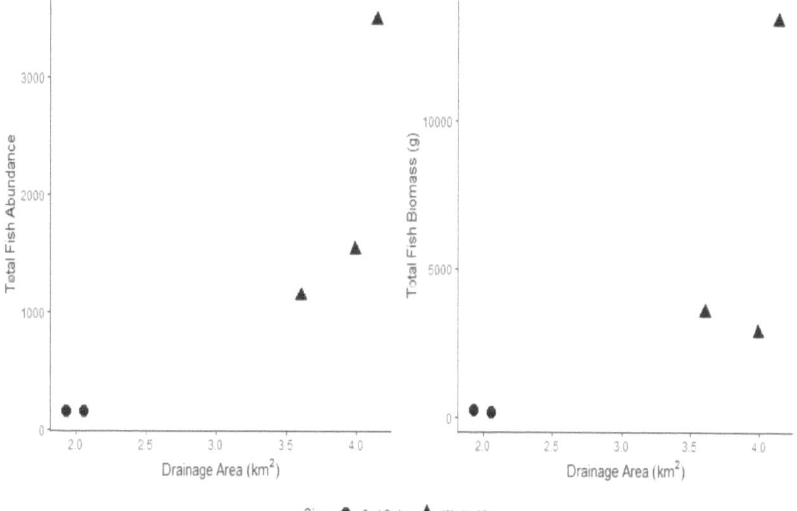

Figure 23

a) AFDM and Total Fish Abundance (Linear: $R^2 = 0.091$, $F_{1,3} = 0.30$, $p = 0.62$) b) AFDM and Total Fish Biomass (Linear: $R^2 = 0.18$, $F_{1,3} = 0.65$, $p = 0.48$)

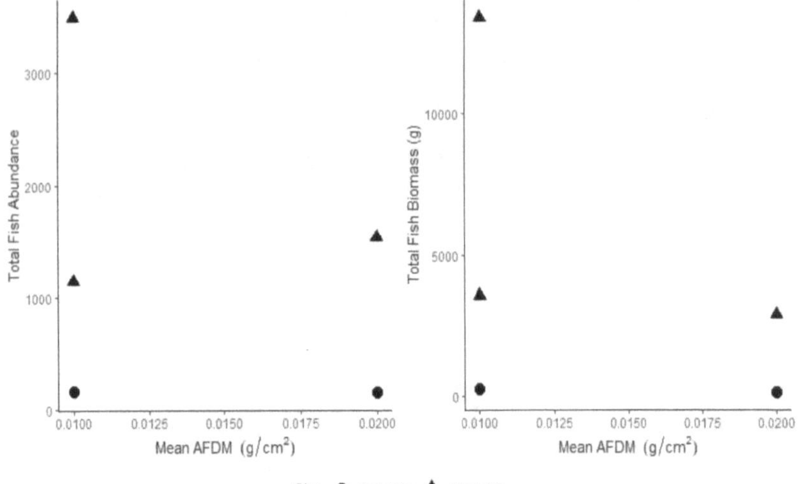

Figure 24

a) Mean Chlorophyll a and Total Fish Abundance (Linear: $R^2 = 0.46$, $F_{1,3} = 2.53$, p = 0.21, Quadratic: $R^2 = 0.64$, $F_{2,2} = 1.75$, p = 0.36, Cubic: $R^2 = 0.97$, $F_{3,1} = 11.04$, p = 0.25). B) Mean Chlorophyll a and Total Fish Biomass (Linear: $R^2 = 0.46$, $F_{1,3} = 3.38$, p = 0.16, Quadratic: $R^2 = 0.78$, $F_{2,2} = 3.53$, p = 0.22, Cubic: $R^2 = 0.95$, $F_{3,1} = 6.74$, p = 0.27)

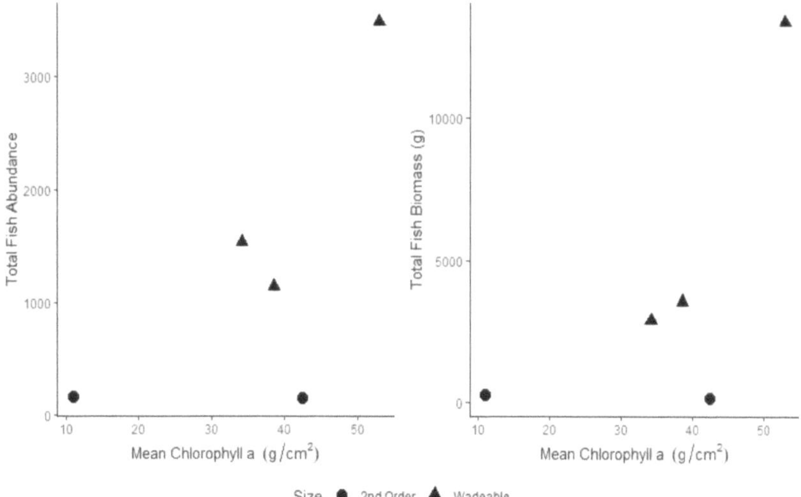

Figure 25

Total Insectivore Biomass and Total Macroinvertebrate Biomass (Linear: R^2 = 0.12, $F_{1,3}$ = 0.42, p = 0.56, Quadratic: R^2 = 0.26, $F_{2,2}$ = 0.35, p = 0.74, Cubic: R^2 = 0.87, $F_{3,1}$ = 2.6, p = 0.42).

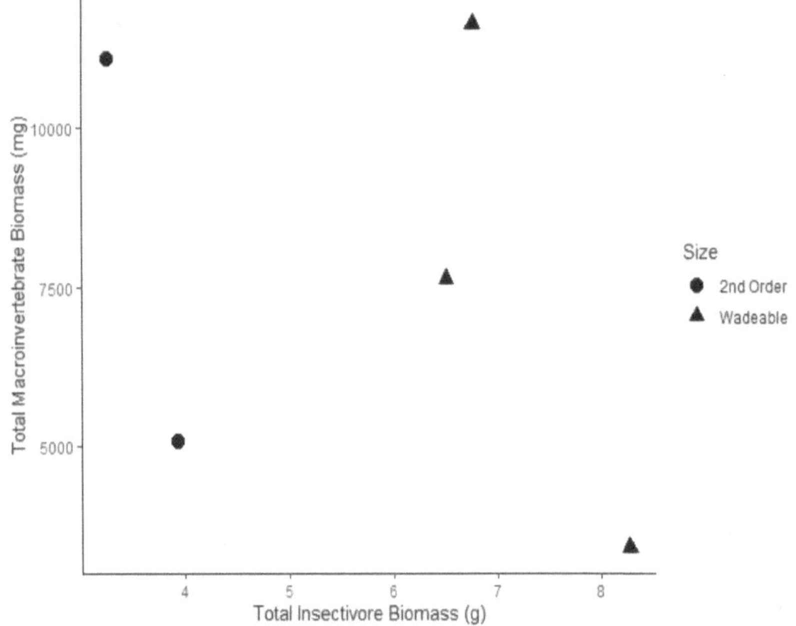

Fish species were categorized according to their trophic designations of herbivores, omnivores, and insectivores. There was significant correlation between any of the feeding groups and drainage area, AFDM, and chlorophyll *a* (Table 6) based on linear regressions. While not significant, there appeared to be a trend toward an overall increase in herbivore, omnivore, and insectivore abundance and biomass (Figures 26 and 28) with increasing drainage area and chlorophyll *a*.

Table 6

Linear Regressions between Herbivore, Omnivore, and Insectivore Abundance and Biomass and Drainage Area, AFDM, and Chlorophyll a.

Regression	R^2	$F_{1,8}$	P value
Herbivore abundance ~ drainage area	0.72	7.74	0.069
Herbivore abundance ~ AFDM	0.004	0.012	0.92
Herbivore abundance ~ chlorophyll *a*	0.46	2.51	0.21
Herbivore biomass ~ drainage area	0.58	4.16	0.77
Herbivore biomass ~ AFDM	$4.83*10^{-5}$	0.00014	0.99
Herbivore biomass ~ chlorophyll *a*	0.46	2.56	0.21
Omnivore abundance ~ drainage area	0.49	2.93	0.19
Omnivore abundance ~ AFDM	0.0025	0.0076	0.94
Omnivore abundance ~ chlorophyll *a*	0.44	2.38	0.22
Omnivore biomass ~ drainage area	0.36	1.71	0.28
Omnivore biomass ~ AFDM	0.014	0.04	0.85
Omnivore biomass ~ chlorophyll *a*	0.43	2.22	0.23
Insectivore abundance ~ drainage area	0.75	8.77	0.059
Insectivore abundance ~ AFDM	0.041	0.13	0.74
Insectivore abundance ~ chlorophyll *a*	0.41	2.12	0.24
Insectivore biomass ~ drainage area	0.49	2.93	0.19
Insectivore biomass ~ AFDM	0.001	0.003	0.96
Insectivore biomass ~ chlorophyll *a*	0.44	2.35	0.22

Figure 26

a) Drainage Area and Herbivore Abundance (Linear: $R^2 = 0.72$, $F_{1,3} = 7.74$, $p = 0.069$, Quadratic: $R^2 = 0.87$, $F_{2,2} = 6.81$, $p = 0.13$, Cubic: $R^2 = 0.90$, $F_{3,1} = 2.93$, $p = 0.4$). b) Mean AFDM and Herbivore Abundance (Linear: $R^2 = 0.004$, $F_{1,3} = 0.012$, $p = 0.92$, Quadratic: $R^2 = 0.52$, $F_{2,2} = 1.08$, $p = 0.48$). c) Mean Chlorophyll a and Herbivore Abundance (Linear: $R^2 = 0.46$, $F_{1,3} = 2.51$, $p = 0.21$, Quadratic: $R^2 = 0.53$, $F_{2,2} = 1.69$, $p = 0.37$, Cubic: $R^2 = 0.96$, $F_{3,1} = 9.04$, $p = 0.24$).

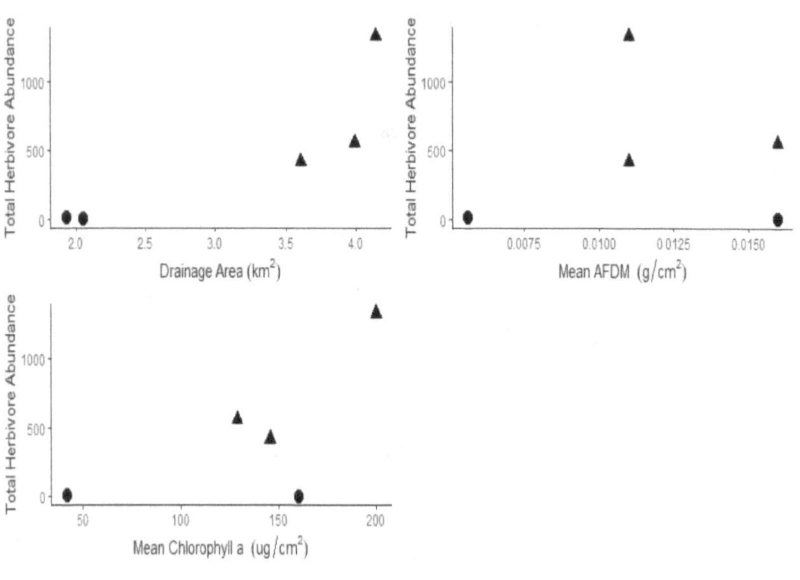

Figure 27

a) Drainage Area and Herbivore Biomass (Linear: $R^2 = 0.58$, $F_{1,3} = 4.16$, $p = 0.13$, Quadratic: $R^2 = 0.77$, $F_{2,2} = 3.45$, $p = 0.22$, Cubic: $R^2 = 0.82$, $F_{3,1} = 1.53$, $p = 0.52$) b) Mean AFDM and Herbivore Biomass (Linear: $R^2 = 4.83*10^{-5}$, $F_{1,3} = 0.00014$, $p = 0.99$, Quadratic: $R^2 = 0.52$, $F_{2,2} = 1.08$, $p = 0.48$) c) Mean Chlorophyll a and Herbivore Biomass (Linear: $R^2 = 0.46$, $F_{1,3} = 2.56$, $p = 0.21$, Quadratic: $R^2 = 0.74$, $F_{2,2} = 2.86$, $p = 0.26$, Cubic: $R^2 = 0.97$, $F_{3,1} = 10.11$, $p = 0.23$)

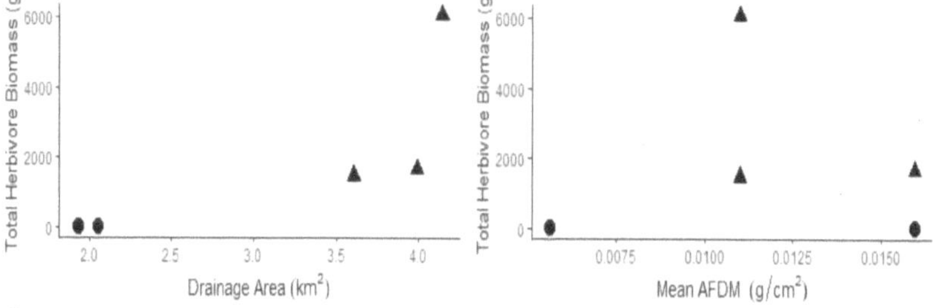

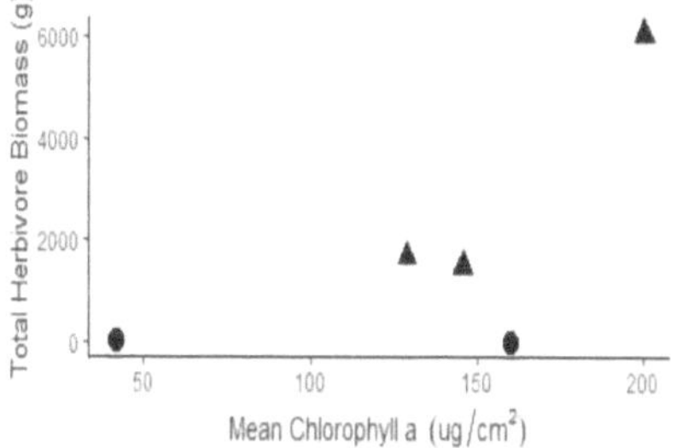

Figure 28

a) Drainage Area and Omnivore Abundance (Linear: $R^2 = 0.49$, $F_{1,3} = 2.93$, $p = 0.19$, Quadratic: $R^2 = 0.72$, $F_{2,2} = 2.61$, $p = 0.28$, Cubic: $R^2 = 0.78$, $F_{3,1} = 1.17$, $p = 0.58$) B) Mean AFDM and Omnivore Abundance (Linear: $R^2 = 0.0025$, $F_{1,3} = 0.0076$, $p = 0.94$, Quadratic: $R^2 = 0.50$, $F_{2,2} = 0.99$, $p = 0.50$) c) Mean Chlorophyll a and Omnivore Abundance (Linear: $R^2 = 0.44$, $F_{1,3} = 2.38$, $p = 0.22$, Quadratic: $R^2 = 0.80$, $F_{2,2} = 3.90$, $p = 0.2$, Cubic: $R^2 = 0.97$, $F_{3,1} = 12.28$, $p = 0.21$)

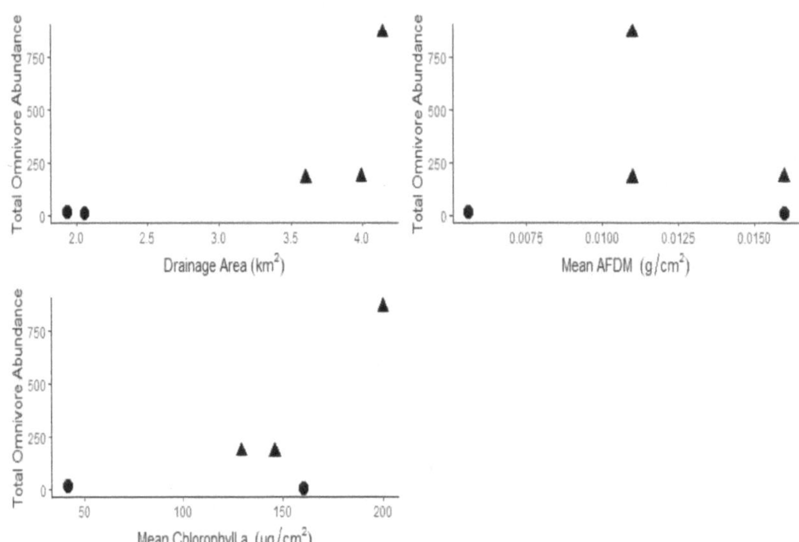

Figure 29

a) Drainage Area and Omnivore Biomass (Linear: $R^2 = 0.36$, $F_{1,3} = 1.71$, $p = 0.28$, Quadratic: $R^2 = 0.60$, $F_{2,2} = 1.52$, $p = 0.40$, Cubic: $R^2 = 0.68$, $F_{3,1} = 0.71$, $p = 0.68$) b) Mean AFDM and Omnivore Biomass (Linear: $R^2 = 0.014$, $F_{1,3} = 0.04$, $p = 0.85$, Quadratic: $R^2 = 0.50$, $F_{2,2} = 0.1.02$, $p = 0.50$) c) Mean Chlorophyll a and Omnivore Biomass (Linear: $R^2 = 0.43$, $F_{1,3} = 2.22$, $p = 0.23$, Quadratic: $R^2 = 0.88$, $F_{2,2} - 7.35$, $p = 0.12$, Cubic: $R^2 = 0.97$, $F_{3,1} = 11.4$, $p = 0.21$)

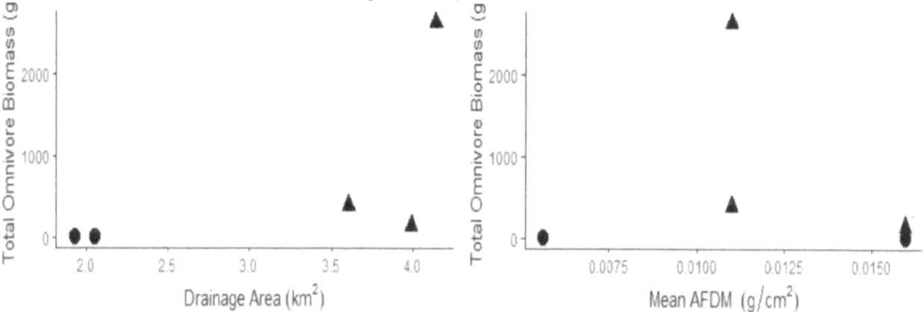

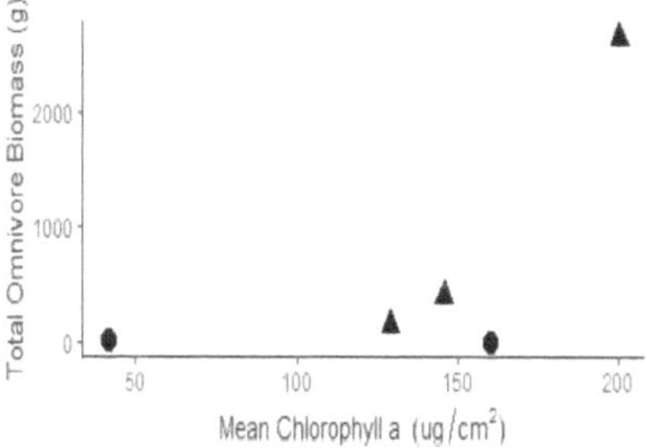

Figure 30

a) *Drainage Area and Insectivore Abundance (Linear: $R^2 = 0.75$, $F_{1,3} = 8.77$, $p = 0.059$, Quadratic: $R^2 = 0.97$, $F_{2,2} = 30.95$, $p = 0.031$, Cubic: $R^2 = 0.99$, $F_{3,1} = 38.95$, $p = 0.12$) b) Mean AFDM and Insectivore Abundance (Linear: $R^2 = 0.041$, $F_{1,3} = 0.13$, $p = 0.74$, Quadratic: $R^2 = 0.36$, $F_{2,2} = 0.55$, $p = 0.64$) c) Mean Chlorophyll a and Insectivore Abundance (Linear: $R^2 = 0.41$, $F_{1,3} = 2.12$, $p = 0.24$, Quadratic: $R^2 = 0.52$, $F_{2,2} = 1.07$, $p = 0.48$, Cubic: $R^2 = 1$, $F_{3,1} = 5.51*10^5$, $p = 0.00099$)*

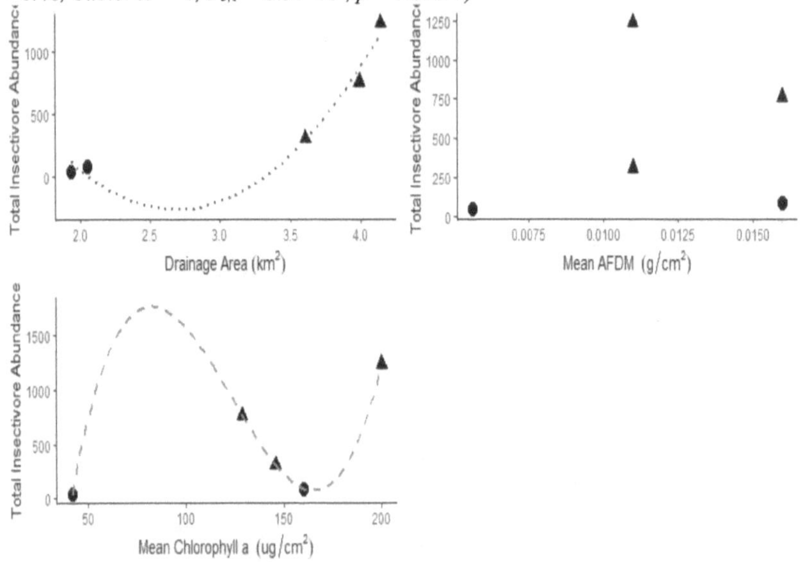

Figure 31

a) Drainage Area and Insectivore Biomass (Linear: $R^2 = 0.49$, $F_{1,3} = 2.93$, $p = 0.19$, Quadratic: $R^2 = 0.75$, $F_{2,2} = 2.99$, $p = 0.25$, Cubic: $R^2 = 0.81$, $F_{3,1} = 1.38$, $p = 0.54$) b) Mean AFDM and Insectivore Biomass (Linear: $R^2 = 0.001$, $F_{1,3} = 0.003$, $p = 0.96$, Quadratic: $R^2 = 0.46$, $F_{2,2} = 0.86$, $p = 0.54$) c) Mean Chlorophyll a and Insectivore Biomass (Linear: $R^2 = 0.44$, $F_{1,3} = 2.35$, $p = 0.22$, Quadratic: $R^2 = 0.79$, $F_{2,2} = 3.79$, $p – 0.21$, Cubic: $R^2 = 0.98$, $F_{3,1} = 20.79$, $p = 0.16$)

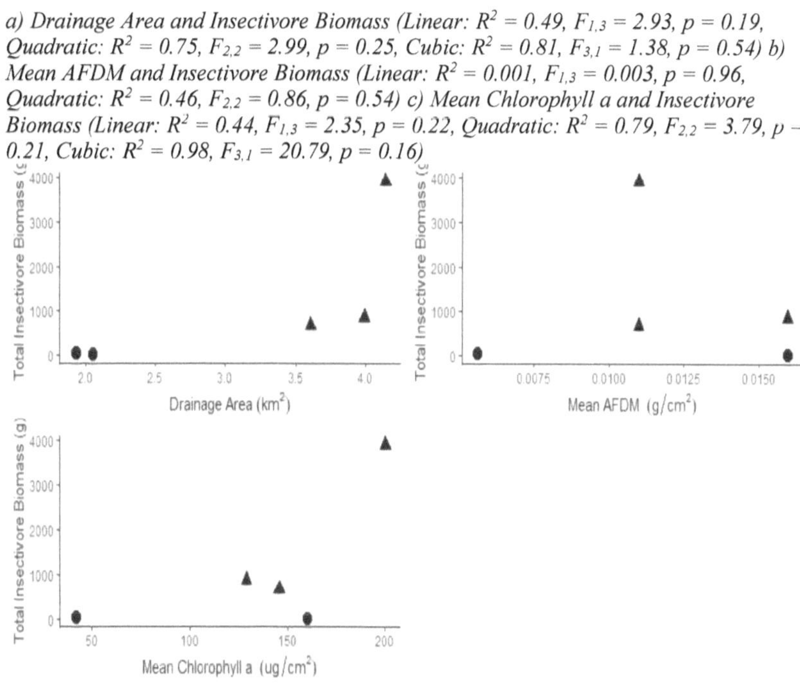

SEM Model

Two structural equation models were analyzed in this study: the original model (Figure 32) and the "best fit" model (Figure 33). All numeric variables were scaled and passed multivariate normality. Neither model passes good model fit parameters, however, the alternative model is the better fit (χ^2 (df = 14, n = 10) = 22.61, p > 0.05, TLI = 0.756, CFI = 0.878) than the original model (χ^2 (df = 5, n = 10) = 21.78, p < 0.05, TLI = -0.329, CFI = 0.76)

Figure 32

Original SEM Model

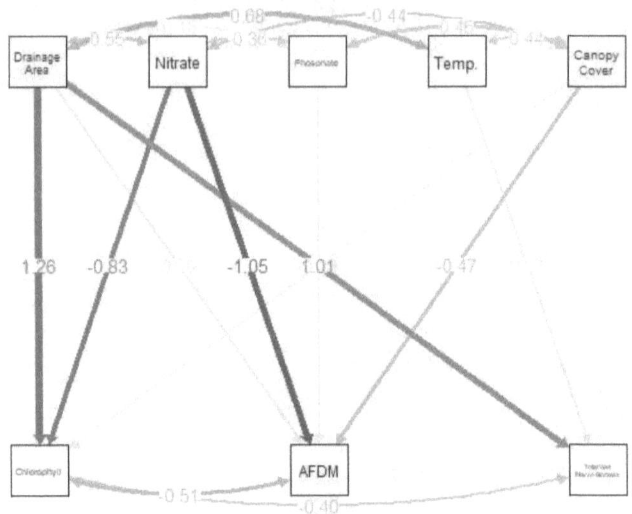

Figure 33

Alternative SEM Model

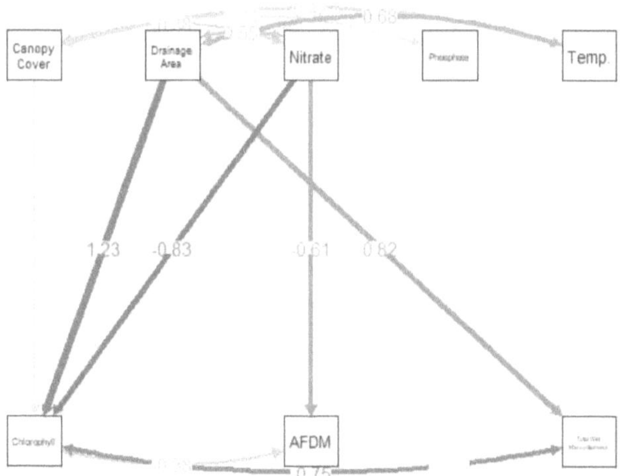

In the alternative model, several relationships were removed due to insignificance. Canopy cover, drainage area, nitrate, and phosphate were regressed against chlorophyll *a*, nitrate against AFDM, and drainage area against total wet macroinvertebrate biomass. Double arrows represent covariances between variables. The SEM model showed that drainage area had a strong impact on chlorophyll *a* (coef. = 0.794, p <0.001) and total wet macroinvertebrate biomass (coef. = 0.685, p <0.001). Nitrate had a strong negative impact on chlorophyll *a* (coef. = -0.779, p <0.001) and AFDM (coef. = -0.592, p = 0.016). There was also a strong covariance between drainage area and temperature (coef.

= 0.877, p = 0.027) and nitrate (coef. = 0.782, p = 0.037). There is no significant

relationship between AFDM and chlorophyll *a* (coef. = -0.041, p = 0.267).

Chapter 4: Discussion

The purpose of this study was to investigate the potential of nutrient release post-restoration on biological communities of periphyton, macroinvertebrates, and fish. The study considered drainage area and water chemistry variables as other possible factors to the responses of biological communities. Many studies have looked at biological responses of nutrient enrichment (Fuller et al. 2008, Miltner 2010, Taylor and Scott 2020), but few consider stream size as a covarying factor to the response. This study compared the biological communities of 4 pre-restoration sites and 6 post-restoration sites of varying stream size and nutrient levels. While this study predicted high nutrient levels post-restoration, there was little difference in nutrient levels between to pre- and post-restoration sites and stream continuum. Rather, chlorophyll *a* increased with an increase in stream size as opposed to restoration status, and macroinvertebrate and fish biomass followed patterns of chlorophyll *a* and stream size with little influence from restoration status.

Water Chemistry

The water chemistry variables varied throughout the stream sizes. Conductivity and pH stayed relatively consistent over the sites, while, as expected, temperature increased with increasing stream size and open canopy. These results are similar to several studies (Fuller et al. 2008, Sabater et al. 2000) that describe the increase in temperature with an increase in sunlight exposure or open canopy. Sabater et al. found a strong longitudinal component with the water temperature (2000). In this study, as the stream increased in drainage area, the canopy cover decreased, and temperatures increased longitudinally. Water pH and conductivity varied only slightly among the sites, and typically stayed within range of pH found in Pennsylvania streams (6.0 – 9.0) and conductivity (150 – 800 µS/cm) to support freshwater fish communities (Huron River Watershed Council 2013). Conductivity in the pre-restoration sites was a little higher (394 – 637 µS/cm) than the post-restoration sites (386 – 534 µS/cm). pH was also consistent over the sites, ranging from 7.5 to 8.5 at the pre-restoration sites and from 7.4 to 8.6 at the post-restoration sites. There was very little variation in pH and conductivity among our study stream sites.

Total nitrate and phosphate levels were expected to increase in post-restoration sites as well as with increasing stream size. Nitrate and phosphate levels varied among stream size, particularly within the smaller, 2nd order streams. There did not appear to be higher N and P levels in post-restoration streams nor with increasing stream size; both nitrate and phosphate levels decreased with an increase in drainage area. The trend of decreasing N and P down the stream continuum was accompanied by increasing chlorophyll *a* levels, total macroinvertebrate biomass, and total fish biomass, suggesting that the nutrients are being quickly assimilated up the food chain (Gafner and Robinson 2007).

In this study, we used a Hach colorimeter onsite, which measured reactive orthophosphate and nitrate using test kit reagents. The Voinovich School research team sent water samples to the DEP Bureau of Laboratories, where the samples were also measured for total nitrate and total phosphorus. The range of samples measured at the laboratory indicated lower nutrient concentrations in July 2020, falling under 1.00 mg/L for both total nitrate and phosphorus. It may be possible that the differences in nitrate and phosphate levels may be a result from differing sampling techniques or differences in stream discharges on the sample days. Water chemistry from high-, mid-, and low-flow hydrological states can differ substantially.

AFDM

Ash-free dry mass is a measure of biofilm biomass derived from subtracting the leftover sand and silt from the total sample weight to obtain a measure of the total organic matter, containing both heterotrophs and autotrophs (Dodds and Whiles 2010). According to our linear regressions, AFDM of the biofilms on rocky substrate at our study site did not follow any pattern pertaining to the water chemistry, drainage area, or chlorophyll *a*. There was no significant relationship between any of the water chemistry variables or changes along the drainage area gradient, unlike chlorophyll *a*. However, the SEM model did show that nitrate negatively correlated with AFDM, opposite of what was expected. Typically, in response to nutrient enrichment events of nitrate and phosphate, AFDM increases (Sharifi and Ghafori 2005) and tends to correlate positively with chlorophyll *a* (Atazadeh et al 2009). However, some studies do not find these relationships. Sabater et al. (2000) found there were no relationship between AFDM and any environmental variables, including nutrient levels in a disturbed small river watershed in Spain. Taylor et al. (2020) reported no response in mean benthic AFDM with phosphorus enrichment, while chlorophyll *a* exhibited a stronger response to the enrichment in a mesocosm experiment nearby Arkansas agricultural watersheds. AFDM is often more strongly correlated with heterotrophic microbes and carbon availability. Heterotrophs tend to obtain energy from dissolved organic material and oxidize organic carbon (Dodds and Whiles 2010). Variation in dissolved carbon might correlate more with AFDM at our sites, but we did not measure carbon. AFDM is also a potential food source for macroinvertebrates. Although the response was not significant, AFDM did decrease slightly with an increase in macroinvertebrate abundance. Fuller et al. (2008) found a 4-fold increase in AFDM in open canopy sites, following a significant decrease in AFDM as a result an increase in macroinvertebrate densities. High grazing pressure at these sites could reduce AFDM, however, there was no decrease in chlorophyll *a*, which would also be expected to decline in response to grazing pressure as well.

Chlorophyll *a*

Chlorophyll *a* levels were expected to increase in response to nutrient enrichment and an increase in stream size. Chlorophyll *a* levels did increase with a rise drainage area, however, levels did not increase along a nutrient gradient as expected. There was instead a negative relationship between chlorophyll *a* and total N and no relationship with P. Both the stepwise regression and the SEM model displayed a positive increase in chlorophyll *a* with drainage area, and a negative relationship with nitrate. Several studies have found that chlorophyll *a* levels generally increase with more nutrients (Miltner 2010, Sabater et al. 2000, Sharifi and Ghafori 2005, Taylor et al. 2020), but the opposite was found in this study. It is possible that high biological productivity during the midsummer sampling season removed nutrients from the water column (Paisley et al. 2011), resulting in higher algal biomass and lower nutrients.

While nutrient levels and chlorophyll *a* did not seem to relate, drainage area did have a strong influence on chlorophyll. Miltner (2010) incorporated drainage area into a study relating chemical and biological variables in Ohio streams and found a positive, significant relationship between dissolved inorganic nitrogen (DIN), total phosphorus, and chlorophyll as well as drainage area. DIN and canopy cover explained most of the variation in chlorophyll *a*, while open canopy, temperature, drainage area, and nitrate explained most of the variation in chlorophyll *a* in ours based in a stepwise regression. As canopy cover opens up following an increase in stream size, the light availability to the stream channel increases. Productivity is expected to higher in mid-order stream sizes compared to headwaters as the canopy opens (Vannote et al. 1980). As light becomes more available, temperatures increase, and translates to an increase in energy and uptake of nutrients such as nitrate (Dodds and Whiles 2010). As predicted, there were strong relationships between an increase In canopy cover, drainage area, temperatures, and chlorophyll *a*, contrasting to a decrease in nitrate.

Seasonal variation may also play a role in the chlorophyll *a* response. We were unable to include seasonal variation in this study due to covid-19 travel restrictions during the time. Many studies incorporate seasonal differences into their study to interpret the patterns of nutrient levels, periphyton biomass, and macroinvertebrate response overtime. Over just 7 days, Fuller et al. saw a 4-fold increase in both AFDM and chlorophyll *a* biomass followed by a suppression with an increase in Chironomidae and Baetidae (2008). Eckert et al. studied algal biomass proliferation under ambient and shaded light conditions in Piedmont headwater streams and found varying responses in winter and spring seasons (2019). In the winter, the algal biomass was highest under high nutrient, ambient light conditions, while in the spring, algal biomass was highest under high nutrient, shaded light conditions. Algal biomass sampling was only conducted once during the summer season for this study. The current trends of the streams described in this study may not depict the typical seasonal patterns of the study system; nutrient levels and periphyton biomass may vary over the season.

Macroinvertebrates

Macroinvertebrate abundance, richness, biomass, evenness and MAIS scores tended to increase with an increase in drainage area. While MAIS scores were low in the anastomosing stream-wetland complexes, they increased in both pre-restoration and post-restoration 2nd order streams and wadeable sites. The low scores in the wetland-stream complexes is most likely due to differences in instream habitat for macroinvertebrates The habitat in the anastomosing streams tended to be more homogenous, with very few riffles and erosional habitat present. At Big Spring Run, MAIS scores showed no significant changes following restoration, contrary to what was expected (Smith et al. 2020). Smith et al. stated that improvement of hydrogeomorphic function may have been at the cost of suitable habitat for macroinvertebrates (2020). Stream size in this study seemed to play a more important role in the response of macroinvertebrates compared to other variables considered.

One macroinvertebrate community metric of interest between post- and pre-restoration sites was % EPT, which decreased following restoration. This also occurred at Big Spring Run, where EPT taxa decreased following restoration. EPT taxa require clean substrate and high oxygen levels, and disturbance following restoration and an increase in suspended sediments that can disrupt sensitive taxa by clogging filter-feeding structures (Big Spring Run Research Results 2019). % EPT taxa commonly decrease with increased nutrients (Justus et al. 2010, Miltner and Rankin 1998, Wang and Garrison 2007). In this study, the % EPT taxa was the highest in the pre-restoration, 2nd order streams, which coincidentally had high nitrate levels at all three 2nd order, pre-restoration sites during the time of sampling (1.5-3.3 mg/L). Thus, at our study sites, nitrate levels may not have as strong of an impact on EPT taxa compared to other studies that have found the opposite response.

Macroinvertebrate abundance and biomass varied across the sites, with lower abundance and biomass in the smaller streams, and higher abundance and biomass in the larger, wadeable streams. Macroinvertebrate abundance and biomass can both give insight into the dynamics of the food web structure of stream ecosystems, and whether the system is controlled by bottom-up or top-down effects. Macroinvertebrate biomass and abundance can also give insight into where nutrients flow in the food web and how macroinvertebrates influence algal and fish biomass. Chlorophyll *a* and AFDM served as a predictor in linear and stepwise regressions against macroinvertebrate total wet biomass and abundance. AFDM did not have any impact on macroinvertebrate abundance or biomass. Macroinvertebrate biomass and abundance increased with the increase in drainage area, as shown in the linear and stepwise regressions. The SEM model as portrayed a significant relationship with drainage area and total macroinvertebrate wet biomass. Macroinvertebrate abundance also increased with an increase in chlorophyll *a*. As drainage area increases, chlorophyll *a* and primary productivity also increases, as predicted by the RCC (Vannote et al. 1980). Periphyton serves as an important food source for macroinvertebrate communities, and in mid-order reaches, periphyton biomass can peak and elicit a positive response in biomass and abundance of macroinvertebrates (Grubaugh et al. 1996).

When macroinvertebrates were categorized into functional feeding groups, however, the CG, CF, and SC feeding groups did not follow as similar of trends as the overall abundance and biomass did. Wets weights of Chironomidae (CG),

Hydropscyhidae (CF), and Elmidae and Heptageniidae (SC), were obtained as a measure of biomass for the feeding groups. Scraper-grazer abundance and biomass did not follow the predictions as strongly as expected; linear regressions showed that SC abundance increased with chlorophyll a as predicted, however, wet biomass did not. The increase in SC abundance in response to an increasing chlorophyll a gradient was expected and has been found in several studies (Eckert et al 2019, Milter and Rankin 1998, Biggs et al. 2000). Periphyton is an important food source in the diet of scraper-grazers, and it is generally the highest in mid-order stream reaches as canopy opens and light increases for optimal conditions of periphyton growth (Grubaugh et al. 1996). Typically, there is a positive relationship between scraper-grazer-biomass and chlorophyll a and light levels (Gjerløv and Richardson 2010, Zimmerman and Death 2002). In this study, however, we only saw a positive relationship between scraper-grazer abundance and not wet biomass. This may be due to the fact that only two representative scraper-grazer taxa were weighed (Heptageniidae and Elmidae), or it could be related to greater error associated with wet biomass measurements. Heptageniidae and Elmidae were expected to be some of the most common macroinvertebrates at the sites, as they were common in reported macroinvertebrate data from Molinari Trib, Molinari, Beham, and Lebanik in 2018-2019. However, Heptageniidae did not turn out to be as common at the sites as predicted. Measuring wet biomass may also have produced some errors, such as missing pieces from the specimens.

Chironomid midges are important in aquatic ecosystems due to their feeding habit and abundance; they are important CG feeders as well as prey for other organisms, as well as tolerant to pollutants and disturbance events. CG abundance did not increase significantly with an increase in drainage area, however, biomass of Chironomidae increased significantly with drainage area. There were no relationships of CG biomass/abundance with chlorophyll *a* or AFDM based on linear regressions. CG abundance did appear to decrease slightly with drainage area and chlorophyll *a* while the biomass increased. It may be possible that there were large Chironomidae in the biomass measures, but few included in the abundance. Chironomids tend to be more tolerant of a wide range of aquatic ecosystems as well as disturbances, which may factor into the increase in biomass with drainage area despite the total CG abundance showing no trend (Fuller et al. 2008). Grubaugh et al. (1996) found an increase in CG biomass along the stream catchment continuum but found varying responses between habitat types. Several studies have described responses in chironomid biomass and abundance with periphyton. Some studies have found positively relations between chironomid and periphyton biomass (Biggs et al. 2000, Eckert et al. 2019), opposite of what was found in this study. Chironomidae feeding can also decrease periphyton biomass (Fuller et al. 2008, Gafner and Robinson 2007). Contrary to these trends, there appeared to be no relationship between CG biomass and periphyton biomass. Some CG, including Chironomidae, tend to feed on the organic compounds in fine sediment, which tends to increase as canopy opens up and the drainage area increases (Vannote et al. 1980). An increase in FPOM along a longitudinal component of area may be one mechanism into the increase of chironomid biomass despite no response to chlorophyll *a*.

Collector-filterers showed the strongest responses with drainage area and chlorophyll *a* based on linear and stepwise regressions. CF biomass (Hydropsychidae) and abundance increased significantly with a response to an increase in chlorophyll *a* as well as drainage area. CF feeding groups are expected to increase downstream as particles become smaller and primary productivity increases (Vannote et al. 1980) but can also be abundant in headwater streams feeding on CPOM and allochthonous detritus (Rosi-Marshall et al. 2016). Eckert et al. (2019) found that CF abundance and biomass increase under shaded light as well as high-nutrient levels. In this study, CF abundance and biomass did increase with increase light levels, contrary to Eckert et al. 2019, but this trend was mainly described by the increase in chlorophyll *a* levels and possibly related to the increase in FPOM. Grubaugh et al. found low CF biomass in headwater cobble riffles, but an increase in CF biomass downstream and with an increase in stream catchment size (1996). As stream size increases, there are depositional areas of sand/silt that collect FPOM. Periphyton biomass also can provide more attachment sites for filtering macroinvertebrates (Grubaugh et al. 1996). CF, including Hydropsychidae, can utilize FPOM and the attachment sites to increase in abundance and biomass, as seen in this study.

One explanation for the lack of positive correlations between food resource biomass (AFDM and chl. *a*) and macroinvertebrate abundance/biomass is that high densities of scraper-grazers can remove AFDM and chl. *a*. There was some evidence of suppression of periphyton growth in this study due to an increase in SC, CG, and CF abundance/biomass. Scraper-grazer abundance does not always change in response to ambient light and periphyton biomass (Gjerløv and Richardson 2010), because scraper-grazer biomass may be related to other water chemistry variables, habitat variables outside of an increase in drainage area, or predation. Molinari, a wadeable, post-restoration site, did have a high biomass of scraper-grazers, contrary to the overall study trend. However, scraper-grazer densities at Molinari and the larger wadeable sites with high chlorophyll *a* levels may not have been high enough to reduce algal biomass (Wallace and Webster 1996). Chironomid and baetid presence may also be an indicator of nutrient enrichment as well as an indication of periphyton suppression (Fuller et al. 2008). Fuller et al. found a significant increase in AFDM and chlorophyll *a* levels at open canopy sites, immediately followed by a decline in biomass in response to an increase in Chironomidae and Baetidae (2019). The increase in the densities of these families elicits a top-down control of periphyton biomass, and while there is some evidence of top-down control of algal growth in our study, the pressure may not have strong enough to seriously deplete chlorophyll *a* levels.

The overall macroinvertebrate community composition differed between the 2nd-order pre-restoration sites and the 2nd order/anastomosing post-restoration sites was best illustrated by the NMDS ordination. The shaded, pre-restoration 2nd order streams (McNay, Poland Run, and Kent) exhibited a high % of EPT taxa, cooler temperatures, and less open canopy compared to the post-restoration streams (Beham, Molinari Trib, Molinari and Lebanik HW). The post-restoration smaller streams were similar in family composition and habitat type, with heavy emergent and overhanging vegetation, and shallow, slow moving water. These sites supported similar family composition to natural/mitigated wetlands in West Virginia (Balcombe et al. 2005). All four of these sites had a high abundance of Physid snails and Turbellaria, or flatworms. All sites except Beham also had a high number of Asellidae, most likely due to Beham having more open water (Balcombe et al. 2005). Asellidae also tend to live in areas of high organic material (detritus) and where fish are not common (Voshell 2002), which may explain why they were not present at Beham. Freshwater snails are mainly scraper-grazers, with some collectors, and have the capability to strongly limit algal biomass by removing the growth from submerged surfaces (McNeely and Power 2007). Flatworms can be found in a variety of habitats, and it has been noted that if there are a high proportion of flatworms at a particular site, it is an indication of organic pollution or nutrient enrichment (Voshell 2002). The phosphate level at Beham during the time of sampling was over 3.3 mg/L but did not have high chlorophyll *a* or AFDM biomass. It is possible that the high scraper-grazer abundance of Physid snails present at Beham and the other smaller sites may have been grazing heavily on the algal biomass at these sites, resulting in low biomass, contrary to lower numbers in the larger, wadeable post-restoration sites with a lower proportion of Physid snails and higher chlorophyll *a* biomass.

Fish

Fish were only present in 5 out of the 10 sites sampled. Fish biomass and abundance did not show a strong relationship with drainage area, AFDM, and chlorophyll *a*, despite predictions. However, if one considers that there were no fish at 5 of the smallest sites, then both fish biomass and abundance increased with an increase in drainage area and chlorophyll *a*. Fish community structures can vary along the drainage area continuum. As habitat diversity increases, fish diversity also tends to increase, and different sized habitats and structures, such as riffle-pool-runs, can be niches for a high diversity of fish (Gebrekiros 2016). Fish abundance, biomass, diversity, and richness all increased as drainage area and chlorophyll *a* increased. There were very few large fish at the sites, but they did become more common as drainage area increased. Newall and Magnuson found that drainage area played a strong factor in community composition in

Wisconsin tributaries (1999). Abundance increased with an increase in stream size, and mid-order stream sizes had many mixed species, including creek chubs, blacknose dace, white suckers, and johnny darters, similar to the species composition found at many of our sites. All of the sites exhibited common warmwater fish species, including central stonerollers, creek chubs, white suckers, and bluntnose minnows, all fairly tolerant, however, stonecat madtoms were found at both Molinari and Lebanik. Stonecat madtoms (*Noturus flavus*) are considered intolerant species are typically found in riffles with clean substrate (Rice and Zimmerman 2019), suggesting both Molinari and Lebanik may be free of silt and dirt of bedrock and cobbles post 3-4 years restoration.

There was some evidence of top-down control of fish on macroinvertebrates. Based on a linear regression, macroinvertebrate biomass did decrease slightly with total insectivore biomass, but not significantly. Typically, an increase in fish may result in a decrease in macroinvertebrates and result in proliferation of algal growth (Biggs et al. 2000, Wallace and Webster 1996). In this study, sites where fish abundance was the highest, chlorophyll *a* levels were also high. There is a possibility that the increase in algal growth at the larger wadeable sites was due to an increase in fish abundance and grazing pressure on the macroinvertebrates (Biggs et al. 2020). As previously mentioned, macroinvertebrates may not have been high enough in abundance to deplete periphyton biomass significantly through grazing pressure. While there may have been heavy herbivory pressures at the sites with high levels of chlorophyll *a* from central stonerollers, it may be that insectivore pressure on grazing macroinvertebrates was high as well, leading to a proliferation of algal growth.

When fish were examined relative to their trophic designation, the abundance and biomass of herbivores, omnivores, and insectivores all increased with drainage area, AFDM, and chlorophyll *a*, although not significantly. Both omnivores and herbivores incorporate periphyton into their diets, where an increase in chlorophyll *a* may factor into an increase in these trophic designations. Herbivore/detritivore/smaller fish tend to increase in enriched areas (Justus et al. 2010). While the wadeable sites were not very nutrient enriched, there was high levels of chlorophyll *a,* creating more food resources for fish communities.

One fish species of interest in our study was the central stoneroller, *Campostoma anomalum,* which was the sole herbivore present at all the sites. Stonerollers graze on algal biomass growing on rocks in riffles. Stoneroller abundance and biomass was high at both Molinari and Lebanik, two of the largest post-restoration streams with high levels of chlorophyll *a*. At the unrestored N. Dunkard Fork, central stoneroller abundance was greater than a thousand per the 100 m^2, and the chlorophyll *a* at this site was the highest out of all the sites. Central stoneroller abundance has been found to positively increase with an increase in algal biomass (Justus et al. 2010), and this species can tolerate low oxygen and proliferate in streams with abundant algal biomass, as seen in N. Dunkard Fork.

SEM Model

SEM models were developed for this study to give insight into the biological connections that may be occurring in the study systems. In this study, there were some limitations to the SEM model, included lack of data and only 10 study sites sampled. The original model did not pass good model fit statistical parameters, including a CFI/TLI above 0.90 to 0.95 and a non-significant χ^2 value. However, the alternative model did present a non-significant χ^2 value and had a larger CFI/TFI than the original model.

Relationships were based off typical biological interactions in headwater to wadeable streams (Sargeant et al. 2012). Stream ecosystems can be complex and can be altered following restoration and increases in nutrients. In the original model, drainage area, nitrate, phosphate, temperature, and canopy cover were regressed against chlorophyll a and AFDM. An increase in nutrients was predicted to have a positive impact on AFDM and chlorophyll a and has been found in several studies (Atazadeh et al. 2009, Nelson et al. 2013). Increased light availability can positively impact nutrient levels (Newcomer et al. 2016) and chlorophyll a and AFDM can be higher in open canopy streams (Fuller et al. 2008). Increased periphyton biomass has also been found to positively influence macroinvertebrate biomass (Justus et al. 2010). In the original model, chlorophyll a and AFDM were regressed against total macroinvertebrate biomass to illustrate this relationship. AFDM and chlorophyll a have been found to covary positively as well (Atazadeh et al 2009), which is depicted in the original model. Canopy cover, temperature, and drainage area also tend to covary with each other: as drainage area increases, the canopy opens, resulting in higher light availability and an increase in temperatures (Fuller et al. 2008, Sabater et al. 2000). Drainage area was covaried with nitrate, phosphate, and temperature in the original model, nitrate with phosphate and canopy cover, phosphate with canopy cover, and temperature and canopy cover. Changes in drainage area can also result in changes in macroinvertebrate biomass, particularly in response to an increase in periphyton biomass as drainage area increases and was illustrated in the original model as well with a regression.

Several insignificant regressions and covariances were removed from the original model to improve the model fit. In the alternative model, only canopy cover, drainage area, nitrate, and phosphate were regressed against chlorophyll *a*, with drainage area relating with chlorophyll *a* positively and nitrate negatively. Only nitrate was regressed against AFDM, which impacted AFDM negatively. Chlorophyll a, AFDM, and drainage area were regressed against total macroinvertebrate wet biomass, where only drainage area impacted the biomass positively. Temperature covaried with drainage area and canopy cover positively. Nitrate covaried negatively with nitrate and positively with phosphate, while drainage area correlated positively with temperature and nitrate. Contrary to what was predicted, AFDM and chlorophyll *a* negatively covaried with each other All other covariances were removed.

Chapter 5: Conclusions

Freshwater ecosystems are among the most threatened ecosystems globally. Restoration attempts to stabilize stream banks and stream channels are most likely to fail as they do not treat the stream as a dynamic system. The restoration project in western Pennsylvania aims to restore to pre-settlement anastomosing wetland-stream complexes, allowing the stream to change over time while reconnecting the stream channel and floodplain and promoting hydrologic connectivity. However, exposing the stream channel to a buildup of legacy sediment over the years has the potential for nutrient enrichment events following restoration. We studied the biological patterns of periphyton, macroinvertebrates, and fish, searching for evidence of nutrient enrichment following restoration.

The water chemistry analysis in this study indicated that there was little nutrient enrichment following 3-4 years post-restoration. Some sites, including Beham and Kent, did have high phosphate and nitrate levels, respectively, but there were no trends with restoration status. Both total nitrate and phosphate varied over the sites and did not increase with drainage area as predicted. Nutrients decreased with a decrease in drainage area and were highly variable over the smaller 2nd order and anastomosing sites. Further analysis of nutrient data at these sites would provide additional nutrient data over the seasons at these sites. There is a possibility that nutrient levels were higher in the beginning of June/July, and then were quickly assimilated into the food chain following an increase in temperatures and sunlight.

Our results also suggest that total N and P levels may not be the best indicator of AFDM and chlorophyll a levels, despite what is sometimes reported by prior studies. Our results suggest that several water chemistry, habitat, and watershed variables may interact to impact the biological responses of periphyton, macroinvertebrates, and fish. There are several mechanisms in aquatic ecosystems that can influence responses of biological communities. In this study, much of the variation was explained by drainage area and chlorophyll a, but there may also be several other variables that interact to impact biological communities that were not explored in this study. There appeared to be stronger evidence of a drainage area gradient impact compared to a nutrient impact on chlorophyll a, AFDM, macroinvertebrate and fish biomass and abundance. Chlorophyll a seemed to be more influenced by drainage area and light availability than nutrients. Macroinvertebrate abundance and biomass was strongly influenced by drainage area, while SC abundance followed trends of chlorophyll a. Fish biomass and abundance followed expected patterns with an increase patterns with stream size, however, the high number of central stonerollers at N. Dunkark Fork was unexpected and seemed to reflect chlorophyll a levels.

To conclude, while there does not appear to be evidence of detrimental levels of nutrients following restoration, there is evidence to support the predictions of an increase in chlorophyll a, macroinvertebrates scraper-grazer and collector-filterer feeding groups, and herbivorous fish as drainage area increases, as supported by the RCC. The anastomosing wetland-stream complexes also exhibited variable water chemistry as well as interesting macroinvertebrate family composition. Natural-wetland stream complexes are rare, and floodplain restoration to create a pre-restoration anastomosing wetland-stream complex may be hydrologically beneficial but does not seem to impact the biological communities as expected. Continuous evaluation of the wetland-stream complexes can give more insight into the biological communities, as well as possible nutrient enrichment events following restoration.

References

Atazadeh, I., Kelly, M., Sharifi, M. and Beardall, J. (2009). The effects of copper and zinc on biomass and taxonomic composition of algal periphyton communities from the River Gharasou, Western Iran. *Oceanological and Hydrobiological Studies, 38*(3), pp.3-14.

Balcombe, C. K., Anderson, J. T., Fortney, R. H., & Kordek, W. S. (2005). Aquatic macroinvertebrate assemblages in mitigated and natural wetlands. *Hydrobiologia, 541*(1), 175-188.

Big Spring Run Research Results. Pennsylvania Department of Environmental Protection. 19. Nov. 2019.

Biggs, B.J., Francoeur, S.N., Huryn, A.D., Young, R., Arbuckle, C.J. and Townsend, C.R. (2000). Trophic cascades in streams: effects of nutrient enrichment on autotrophic and consumer benthic communities under two different fish predation regimes. *Canadian Journal of Fisheries and Aquatic Sciences, 57*(7), pp.1380-1394.

Chapter, P. C. 93: Water Quality Standards.

Chipps, S.R., Hubbard, D.E., Werlin, K.B., Haugerud, N.J., Powell, K.A., Thompson, J. and Johnson, T. (2006). Association between wetland disturbance and biological attributes in floodplain wetlands. *Wetlands, 26*(2), pp.497-508.

Drerup, S.A. (2016). *Functional Responses of Stream Communities to Acid Mine Drainage Remediation* (Doctoral book, Ohio University).

Dodds, W., and M. Whiles. (2010). *Freshwater ecology: concepts and environmental applications*. Elsevier.

Eckert, R.A., Halvorson, H.M., Kuehn, K.A. and Lamp, W.O. (2020). Macroinvertebrate community patterns in relation to leaf-associated periphyton under contrasting light and nutrient conditions in headwater streams. *Freshwater Biology, 65*(7), pp.1270-1287.

First Pennsylvania Resource. Robinson Fork Mitigation Bank Phase I Project Introductory Package, Washington County, Pennsylvania, 5 Feb. 2014. Canonsburg, Pennsylvania.

Foster, J.R. and Lewis, R., Drainage basin size as a predictor of fish species richness in the Otsego Lake Watershed.

Frey, J.W., Bell, A.H., Hambrook Berkman, J.A. and Lorenz, D.L. (2011). *Assessment of nutrient enrichment by use of algal-, invertebrate-, and fish-community attributes in wadeable streams in ecoregions surrounding the Great Lakes*. United States Geological Survey.

Fuller, R. L., LaFave, C., Anastasi, M., Molina, J., Salcedo, H., & Ward, S. (2008). The role of canopy cover on the recovery of periphyton and macroinvertebrate communities after a month-long flood. *Hydrobiologia, 598*(1), 47-57.

Gafner, K., & Robinson, C. T. (2007). Nutrient enrichment influences the responses of stream macroinvertebrates to disturbance. *Journal of the North American Benthological Society, 26*(1), 92-102.

Gebrekiros, S.T. (2016). Factors affecting stream fish community composition and habitat suitability. Journal of Aquaculture and Marine Biology, 4(2), p.00076.

Genito, D., Gburek, W.J. and Sharpley, A.N. (2002). Response of stream macroinvertebrates to agricultural land cover in a small watershed. *Journal of Freshwater Ecology*, *17*(1), pp.109-119.

Gjerløv, C. and Richardson, J.S. (2010). Experimental increases and reductions of light to streams: effects on periphyton and macroinvertebrate assemblages in a coniferous forest landscape. *Hydrobiologia*, *652*(1), pp.195-206.

Gilkinson, K.D., Gordon Jr, D.C., MacIsaac, K.G., McKeown, D.L., Kenchington, E.L., Bourbonnais, C. and Vass, W.P. (2005). Immediate impacts and recovery trajectories of macrofaunal communities following hydraulic clam dredging on Banquereau, eastern Canada. *ICES Journal of Marine Science*, *62*(5), pp.925-947.

Greenwood, J.L. and Rosemond, A.D. (2005). Periphyton response to long-term nutrient enrichment in a shaded headwater stream. *Canadian Journal of Fisheries and Aquatic Sciences*, *62*(9), pp.2033-2045.

Grubaugh, J. W., Wallace, J. B., & Houston, E. S. (1996). Longitudinal changes of macroinvertebrate communities along an Appalachian stream continuum. *Canadian Journal of Fisheries and Aquatic Sciences*, *53*, 896-909.

Higgins, C.L. (2009). Spatiotemporal variation in functional and taxonomic organization of stream-fish assemblages in central Texas. *Aquatic Ecology*, *43*(4), pp.1133-1141.

Hill, W.R., Boston, H.L. and Steinman, A.D. (1992). Grazers and nutrients simultaneously limit lotic primary productivity. *Canadian Journal of Fisheries and Aquatic Sciences*, *49*(3), pp.504-512.

Huron River Watershed Council. "Conductivity." Sept. 2013.

Johnson, Kelly S. "Field and laboratory methods for using the MAIS (Macroinvertebrate

Aggregated Index for Streams) in rapid bioassessment of Ohio streams." *Athens: Ohio University Department of Biological Sciences* (2007).

Jones, C.N., Scott, D.T., Guth, C., Hester, E.T. and Hession, W.C. (2015). Seasonal variation in floodplain biogeochemical processing in a restored headwater stream. *Environmental Science & Technology*, *49*(22), pp.13190-13198.

Justus, B.G., Petersen, J.C., Femmer, S.R., Davis, J.V. and Wallace, J.E. (2010). A comparison of algal, macroinvertebrate, and fish assemblage indices for assessing low-level nutrient enrichment in wadeable Ozark streams. *Ecological Indicators*, *10*(3), pp.627-638.

Kido, R. R., & Kneitel, J. M. (2021). Eutrophication effects differ among functional groups in vernal pool invertebrate communities. *Hydrobiologia*, *848*(7), 1659-1673.

LandStudies, Inc. Ryerson Station State Park Conceptual Design 15% Design Report. Feb. 2019. Lititz, Pennsylvania.

Liston, S.E., Newman, S. and Trexler, J.C. (2008). Macroinvertebrate community response to eutrophication in an oligotrophic wetland: an in situ mesocosm experiment. *Wetlands*, *28*(3), pp.686-694.

McMillan, S.K. and Noe, G.B. (2017). Increasing floodplain connectivity through urban stream restoration increases nutrient and sediment retention. *Ecological Engineering*, *108*, pp.284-295.

McNeely, C., Finlay, J.C. and Power, M.E. (2007). Grazer traits, competition, and carbon sources to a headwater-stream food web. *Ecology*, *88*(2), pp.391-401.

Miltner, R.J. (2010). A method and rationale for deriving nutrient criteria for small rivers and streams in Ohio. *Environmental management*, *45*(4), pp.842-855.

Miltner, R.J. and Rankin, A.E.T. (1998). Primary nutrients and the biotic integrity of rivers and streams. *Freshwater Biology*, *40*(1), pp.145-158.

Miyasaka, H., Genkai-Kato, M., Miyake, Y., Kishi, D., Katano, I., Doi, H., Ohba, S.Y. and Kuhara, N. (2008). Relationships between length and weight of freshwater macroinvertebrates in Japan. *Limnology*, *9*(1), pp.75-80.

Moore Jr, E.L. (2010). *An index of biotic integrity for macroinvertebrates and salamanders in primary headwater habitat streams in Ohio* (Doctoral book, The Ohio State University).

Nanson, G.C. and Knighton, A.D. (1996). Anabranching rivers: their cause, character and classification. *Earth surface processes and landforms*, *21*(3), pp.217-239.

Nelson, C.E., Bennett, D.M. and Cardinale, B.J. (2013). Consistency and sensitivity of stream periphyton community structural and functional responses to nutrient enrichment. *Ecological Applications*, *23*(1), pp.159-173.

Newall, P.R. and Magnuson, J.J. (1999). The importance of ecoregion versus drainage area on fish distributions in the St. Croix River and its Wisconsin tributaries. *Environmental Biology of Fishes*, *55*(3), pp.245-254.

Newcomer, Johnson, T.A., Kaushal, S.S., Mayer, P.M., Smith, R.M. and Sivirichi, G.M. (2016). Nutrient retention in restored streams and rivers: A global review and synthesis. *Water*, *8*(4), p.116.

Ohio Environmental Protection Agency (Ohio EPA). Volume III Standard Biological Field Sampling and Laboratory Methods for Assessing Fish and Macroinvertebrate Communities. 26 June 2015. Columbus, Ohio

Oksanen, J. (2013). Vegan: ecological diversity. *R Project, 368.*

Olde, Venterink, H., Vermaat, J.E., Pronk, M., Wiegman, F., van der Lee, G.E., van den Hoorn, M.W., Higler, L.W.G. and Verhoeven, J.T. (2006). Importance of sediment deposition and denitrification for nutrient retention in floodplain wetlands. *Applied Vegetation Science, 9*(2), pp.163-174.

PA DEP Final Report. Big Spring Run Natural Floodplain, Stream, and Riparian Wetland-Aquatic Resource Restoration Project Monitoring, 2013.

PA DEP. "Water Quality Monitoring Protocols for Streams and Rivers." Office of Water Programs, Bureau of Clean Water, 2018.

Paillex, A., Dolédec, S., Castella, E. and Mérigoux, S. (2009). Large river floodplain restoration: predicting species richness and trait responses to the restoration of hydrological connectivity. *Journal of Applied Ecology, 46*(1), pp.250-258.

Paisley, M.F., Walley, W.J. and Trigg, D.J. (2011). Identification of macro-invertebrate taxa as indicators of nutrient enrichment in rivers. *Ecological Informatics, 6*(6), pp.399-406.

Poos, M.S., Mandrak, N.E. and McLaughlin, R.L. (2007). The effectiveness of two common sampling methods for assessing imperilled freshwater fishes. *Journal of Fish Biology, 70*(3), pp.691-708.

Qu, X., Peng, W., Liu, Y., Zhang, M., Ren, Z., Wu, N. and Liu, X. (2019). Networks and ordination analyses reveal the stream community structures of fish, macroinvertebrate and benthic algae, and their responses to nutrient enrichment. *Ecological Indicators, 101*, pp.501-511.

Rader, R.B. and Richardson, C.J. (1994). Response of macroinvertebrates and small fish to nutrient enrichment in the northern Everglades. *Wetlands, 14*(2), pp.134-146.

Reisinger, A.J., Presuma, D.L., Gido, K.B. and Dodds, W.K. (2011). Direct and indirect effects of central stoneroller (Campostoma anomalum) on mesocosm recovery following a flood: can macroconsumers affect denitrification? *Journal of the North American Benthological Society, 30*(3), pp.840-852.

Rice, D., and B. Zimmerman. 2019. A Naturalist's Guide to the Fishes of Ohio. Special Publication of the Ohio Biological Survey. vii + 391 p.

Roley, S.S., Tank, J.L., Stephen, M.L., Johnson, L.T., Beaulieu, J.J. and Witter, J.D. (2012). Floodplain restoration enhances denitrification and reach-scale nitrogen removal in an agricultural stream. *Ecological Applications, 22*(1), pp.281-297.

Rosi-Marshall, E. J., Vallis, K. L., Baxter, C. V., & Davis, J. M. (2016). Retesting a prediction of the River Continuum Concept: autochthonous versus allochthonous resources in the diets of invertebrates. *Freshwater Science, 35*(2), 534-543.

Rosseel, Y. (2012). Lavaan: An R package for structural equation modeling and more. Version 0.5–12 (BETA). *Journal of Statistical Software, 48*(2), pp.1-36.

Sabater, S., Armengol, J., Comas, E., Sabater, F., Urrizalqui, I. and Urrutia, I. (2000). Algal biomass in a disturbed Atlantic river: water quality relationships and environmental implications. *Science of the Total Environment, 263*(1-3), pp.185-195.

Salvador, P.G., Bravard, J.P., Vital, J. and Voruz, J.L. (1993). Archaeological evidence for Holocene floodplain development in the Rhône valley, France. *Z. Geomorphology. NF, 88*, pp.81-95.

Sargeant, B.L., Gaiser, E.E. and Trexler, J.C. (2011). Indirect and direct controls of macroinvertebrates and small fish by abiotic factors and trophic interactions in the Florida Everglades. *Freshwater Biology*, *56*(11), pp.2334-2346.

Sharifi, M. and Ghafori, M. (2005). Effects of added nutrients on dry mass, AFDM, chlorophyll a and biovolume of periphyton algae in artificial streams.

Sheldon, A.L. (1968). Species diversity and longitudinal succession in stream fishes. *Ecology*, *49*(2), pp.193-198.

Smith, R.F., Neideigh, E.C., Rittle, A.M. and Wallace, J.R. (2020). Assessing macroinvertebrate community response to restoration of Big Spring Run: Expanded analysis of before-after-control-impact sampling designs. *River Research and Applications*, *36*(1), pp.79-90.

Stevenson, R.J. and Bahls, L.L. (1999). Periphyton protocols. *Rapid bioassessment protocols for use in wadeable streams and rivers: periphyton, benthic macroinvertebrates, and fish. EPA.*

Stoffels, R.J., Clarke, K.R., Rehwinkel, R.A. and McCarthy, B.J. (2014). Response of a floodplain fish community to river-floodplain connectivity: natural versus managed reconnection. *Canadian Journal of Fisheries and Aquatic Sciences*, *71*(2), pp.236-245.

Stoffels, R.J., Karbe, S. and Paterson, R.A. (2003). Length-mass models for some common New Zealand littoral-benthic macroinvertebrates, with a note on within-taxon variability in parameter values among published models. *New Zealand Journal of Marine and Freshwater Research*, *37*(2), pp.449-460.

Taylor, J.M. (2011). *Nutrient enrichment effects on stream periphyton stoichiometry, algal and fish assemblage structure, and grazing fish-periphyton interactions* (Doctoral book).

Taylor, J.M., Rodman, A.R. and Scott, J.T. (2020). *Stream algal biomass response to experimental phosphorus and nitrogen gradients: A case for dual nutrient management in agricultural watersheds* (Vol. 49, No. 1, pp. 140-151).

"The Great Marsh Institute," *WebPress.* https://www.greatmarshinstitute.org/about/

Tiner, R.W. (1987). *Mid-Atlantic wetlands: a disappearing natural treasure.* US Fish and Wildlife Service, Fish and Wildlife Enhancement, National Wetlands Inventory Project.

Tockner, K., Pennetzdorfer, D., Reiner, N., Schiemer, F. and Ward, J.V. (1999). Hydrological connectivity, and the exchange of organic matter and nutrients in a dynamic river–floodplain system (Danube, Austria). *Freshwater Biology, 41*(3), pp.521-535.

U.S. Department of Agriculture: Natural Resource Conservation Service. Web Soil Survey. 2019

U.S Geological Survey, 2012. StreamStats Application. https://streamstats.usgs.gov/ss/

Van Nieuwenhuyse, E.E. and Jones, J.R. (1996). Phosphorus chlorophyll relationship in temperate streams and its variation with stream catchment area. *Canadian Journal of Fisheries and Aquatic Sciences, 53*(1), pp.99-105.

Vannote, R.L., Minshall, G.W., Cummins, K.W., Sedell, J.R. and Cushing, C.E. (1980). The river continuum concept. *Canadian Journal of Fisheries and Aquatic Sciences, 37*(1), pp.130-137.

Voshell, J. R. (2002). *A guide to common freshwater invertebrates of North America* (No.

Sirsi) i9780939923878). McDonald & Woodward Pub.

Wallace, J. B., & Webster, J. R. (1996). The role of macroinvertebrates in stream ecosystem function. *Annual Review of Entomology, 41*(1), 115-139.

Walter, R., Merritts, D. and Rahnis, M. (2007). Estimating volume, nutrient content, and rates of stream bank erosion of legacy sediment in the Piedmont and Valley and Ridge physiographic provinces, southeastern and central PA. *Report to the Pennsylvania Department of Environmental Protection, Lancaster, Pennsylvania.*

Wang, L., Robertson, D.M. and Garrison, P.J. (2007). Linkages between nutrients and assemblages of macroinvertebrates and fish in wadeable streams: implication to nutrient criteria development. *Environmental Management, 39*(2), pp.194-212.

Wetzel, M.A., Leuchs, H. and Koop, J.H. (2005). Preservation effects on wet weight, dry weight, and ash-free dry weight biomass estimates of four common estuarine macro-invertebrates: no difference between ethanol and formalin. *Helgoland Marine Research, 59*(3), pp.206-213.

Wohl, E. and Merritts, D.J. (2007). What is a natural river? *Geography Compass, 1*(4), pp.871-900.

Wolf, K.L., Noe, G.B. and Ahn, C. (2013). Hydrologic connectivity to streams increases nitrogen and phosphorus inputs and cycling in soils of created and natural floodplain wetlands. *Journal of Environmental Quality, 42*(4), pp.1245-1255.

Zimmermann, E. M., & Death, R. G. (2002). Effect of substrate stability and canopy cover on stream invertebrate communities. *New Zealand Journal of Marine and Freshwater Research, 36*(3), 537-545.

Appendix A: Illustrations and Maps of Sites

Figure 34

Pre-restoration Floodplain Covered in Legacy Sediment (Big Spring Run Research Results, 2019)

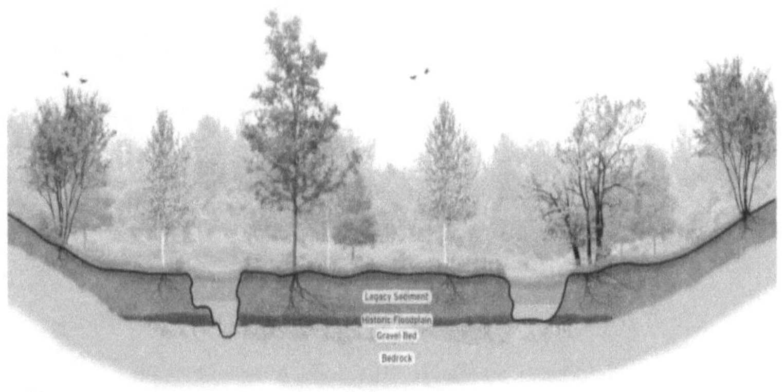

Figure 35

Goal of Floodplain Restoration (Big Spring Run Research Results, 2019)

Figure 36

Map of Sites at Ryerson Station State Park

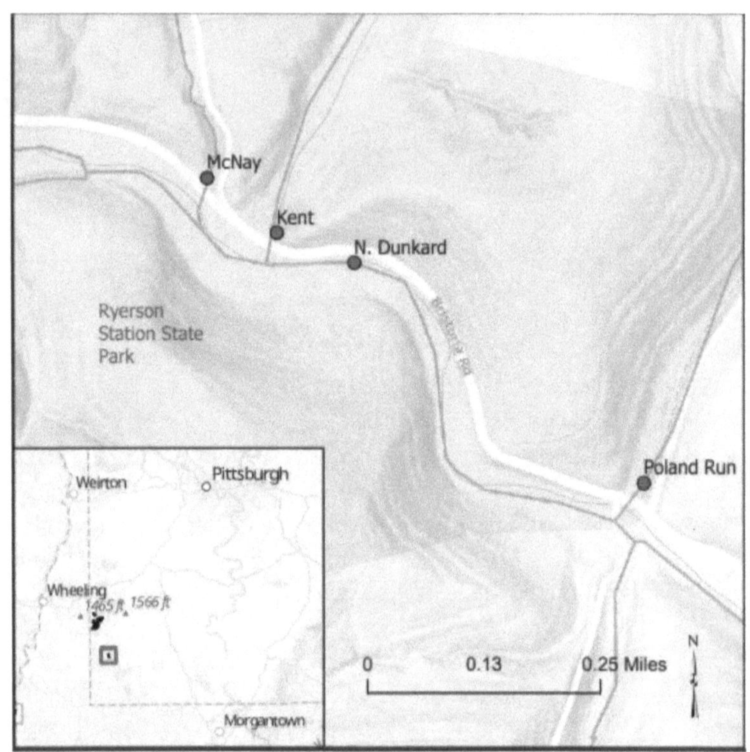

Figure 37

Map of Sites at Robinson Run

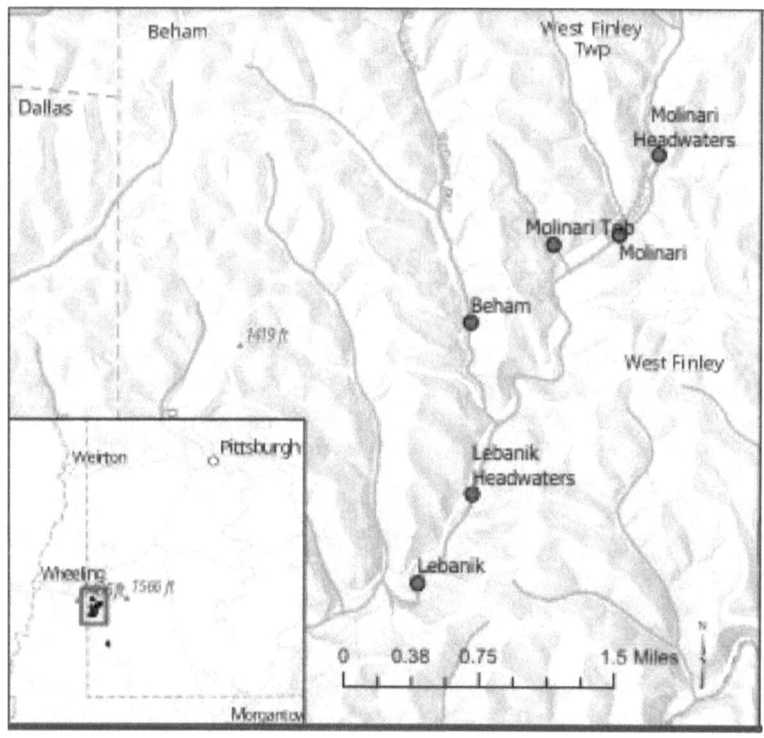

Appendix B: Raw Biological Data

Table 7

AFDM Raw Measurements at each Site

Site	Replicate	AFDM (g/cm^2)
Molinari HW	1	0.00801587
Molinari HW	2	0.00878307
Molinari HW	3	0.01314815
Lebanik HW	1	0.00812169
Lebanik HW	2	0.02201058
Lebanik HW	3	0.01634921
Molinari Trib	1	0.00431217
Molinari Trib	2	0.00058201
Molinari Trib	3	0.00194444
McNay	1	0.005
McNay	2	0.00375661
McNay	3	0.00380952
Poland Run	1	0.01153439
Poland Run	2	0.02005291
Poland Run	3	0.03994709
Kent	1	0.00714286
Kent	2	0.005
Kent	3	0.00473545
Beham	1	0.01214286

Beham	2	0.01830688
Beham	3	0.01724868
Molinari	1	0.01291005
Molinari	2	0.01306878
Molinari	3	0.00579365
Lebanik	1	0.00992063
Lebanik	2	0.01063492
Lebanik	3	0.02706349
N. Dunkard Fork	1	0.01071429
N. Dunkard Fork	2	0.00965608
N. Dunkard Fork	3	0.01291005

Table 8

Chlorophyll a Raw Measurements at each Site

Site	Replicate	Chlorophyll a ($\mu g/cm^2$)	Site	Replicate	Chlorophyll a ($\mu g/cm^2$)
Lebanik HW	#1A	8.957549858	Kent	#3C	22.96784434
Lebanik HW	#1B	19.15715716	Kent	#1A	16.58024691
Lebanik HW	#1C	9.551010718	Kent	#1C	20.93336193
Lebanik HW	#2A	16.28725926	Kent	#2A	12.96875
Lebanik HW	#2B	16.17248677	Kent	#2B	10.375
Lebanik HW	#2C	18.60705467	Kent	#2C	12.80864198
Lebanik HW	#3A	8.571075838	Kent	#3A	0.593703704
Lebanik HW	#3B	5.85708061	Kent	#3B	0.645925926
Lebanik HW	#3C	7.984555985	Kent	#3C	0.674074074
Molinari HW	#1A	12.94942421	Beham	#1A	24.53520075
Molinari HW	#1B	11.94412182	Beham	#1B	34.95238095

Molinari HW	#1C	2.514430014	Beham	#1C	31.53481481
Molinari HW	#2A	2.353451458	Beham	#2A	47.22439852
Molinari HW	#2B	1.952380952	Beham	#2B	49.75252205
Molinari HW	#2C	18.32424829	Beham	#2C	72.10893246
Molinari HW	#3A	21.50289562	Beham	#3A	36.6968254
Molinari HW	#3B	18.94711199	Beham	#3B	38.61351052
McNay Run	#3C	3.272751323	Beham	#3C	46.23628118
McNay Run	#1A	4.305723906	Molinari	#1A	44.64401544
McNay Run	#1B	7.455137845	Molinari	#1B	53.73755102
McNay Run	#1C	1.283421517	Molinari	#1C	41.72272727
McNay Run	#2A	1.215873016	Molinari	#2A	35.89618683
McNay Run	#2B	1.204293273	Molinari	#2B	54.31804821
McNay Run	#2C	1.019269103	Molinari	#2C	44.7034632
McNay Run	#3A	0.517420635	Molinari	#3A	24.79532164
McNay Run	#3B	1.205162738	Molinari	#3B	22.94485597

Poland Run	#3C	11.98931624	Molinari	#3C	24.54234234
Poland Run	#1A	11.97901235	Lebanik	#1A	28.12208217
Poland Run	#1B	8.019097222	Lebanik	#1B	25.35972384
Poland Run	#1C	27.68699924	Lebanik	#1C	29.78174603
Poland Run	#2A	24.93514739	Lebanik	#2A	39.72270813
Poland Run	#2B	34.88262911	Lebanik	#2B	35.00916306
Poland Run	#2C	22.25672878	Lebanik	#2C	41.98641975
Poland Run	#3A	22.53256003	Lebanik	#3A	35.76719577
Poland Run	#3B	22.86358257	Lebanik	#3B	37.5308642
Molinari Trib	#3C	1.378654971	Lebanik	#3C	34.13164021
Molinari Trib	#1A	1.587936508	N. Dunkard Fork	#1A	45.6036556
Molinari Trib	#1B	1.105144033	N. Dunkard Fork	#1B	48.52280852
Molinari Trib	#1C	0.400992063	N. Dunkard Fork	#1C	46.71143671
Molinari Trib	#2A	0.840460629	N. Dunkard Fork	#2A	31.06603175
Molinari Trib	#2B	0.715041572	N. Dunkard Fork	#2B	41.67375887

Molinari Trib	#2C	0.803915344	N. Dunkard Fork	#2C	38.80808081
Molinari Trib	#3A	0.833444075	N. Dunkard Fork	#3A	51.76900585
Molinari Trib	#3B	0.638624339	N. Dunkard Fork	#3B	92.5751634
			N. Dunkard Fork	#3C	79.93791887

Table 9

Macroinvertebrate Family and Feeding Group Abundance at each Site

Family	Feeding Group	Lebanik HW	Molinari HW	McNay Run	Poland Run	Molinari Trib	Kent Run	Beham	Molinari	Lebanik	N. Dunkard Fork
Aeshnidae	PR	4	0	0	1	1	5	3	11	2	0
Asellidae	CG	101	113	2	0	198	0	0	9	0	0
Ameltidae		0	0	0	0	0	0	0	0	0	1
Athericidae	PR	0	0	0	0	0	2	0	6	3	5
Baetidae	CG	0	0	0	6	0	1	0	8	3	0
Caenidae	CG	0	0	0	2	1	1	1	57	20	7
Calopterygidae	PR	0	0	0	0	1	0	0	2	2	0
Cambaridae	GN	0	0	3	1	0	3	0	19	15	9
Capniidae	SH	0	2	0	0	0	153	0	0	0	0

Chironomidae	CG	0	4	2	160	2	19	10	16	68	49
Chloroperlidae	PR	0	0	0	75	0	0	0	4	2	0
Corixidae	MP	9	0	0	0	0	0	0	0	14	0
Corydalidae	PR	0	0	2	2	0	3	0	1	1	0
Dryopidae	SC	0	0	0	3	1	1	2	9	2	0
Dytiscidae	PR	17	0	3	1	0	1	0	3	5	0
Elmidae	SC	0	6	0	18	0	14	77	120	44	76
Ephemerellidae	CG	0	0	0	1	0	0	0	2	6	1
Ephemeridae	CG	0	0	1	15	0	3	0	8	17	9
Gammaridae	CG	0	0	0	0	0	0	0	0	0	3
Gerridae	PR	0	0	0	0	0	0	0	0	1	0
Gomphidae	PR	0	0	0	1	0	0	0	1	0	1
Haliplidae	MP	12	0	0	0	0	0	0	2	0	0
Helicopsychidae	SC	0	0	0	0	0	0	4	0	11	0
Heptageniidae	SC	0	0	0	5	0	3	0	16	5	0
Hydrophilidae	PR	0	0	0	0	1	0	1	0	5	0

Hydropsychidae	CF	0	1	5	19	0	12	6	67	60	158
Hydroptilidae	MP	0	0	0	0	0	0	0	0	0	1
Isonychiidae	CF	0	0	0	0	0	0	0	4	4	6
Lestidae	PR	0	0	0	0	0	0	0	0	3	0
Leuctridae	SH	0	4	0	28	0	65	3	6	0	2
Libellulidae	PR	0	0	1	0	0	0	0	0	0	0
Limnephilidae	SH	0	0	0	0	1	3	5	3	0	0
Limoniidae		0	0	0	4	0	18	0	0	1	7
Lymnaeidae	CG	0	0	0	0	0	0	5	0	0	0
Notonectidae	PR	1	0	18	0	0	1	0	6	24	2
Peltoperlidae	SH	0	0	4	0	0	0	0	0	0	0
Perlidae	PR	0	2	29	2	0	10	9	14	9	4
Perlodidae	PR	0	0	7	0	0	0	1	0	4	0
Philopotamidae	CF	0	0	0	2	1	0	0	0	0	0
Physidae	SC	156	61	31	1	62	3	180	90	122	9
Planorbidae	CG	0	0	0	0	0	0	0	0	1	0

Psephenidae	SC	0	0	0	2	0	1	3	44	14	28
Rhyacophilidae	PR	0	0	0	1	0	1	0	0	0	0
Sialidae	PR	0	0	0	0	0	0	0	0	1	0
Sphaeridae	CF	1	0	0	0	0	1	2	3	2	0
Tabanidae	PR	0	0	0	0	0	0	4	0	0	0
Tipulidae	SH	0	0	1	3	0	10	1	0	2	1
Turbellaria	CG	0	79	2	0	82	0	148	2	0	0
Veliidae	PR	1	0	0	8	0	4	1	3	3	0

Table 10

Fish Species and Trophic Designation at each Site

Species	Trophic Designation	Molinari	Lebanik	Beham	Kent	N. Dunkard
Stoneroller minnow	Herbivore	421	557	5	14	1330
Creek chub	Generalist	190	23	45	85	52
White sucker	Omnivore	127	118	5	14	125
Bluegill sunfish	Generalist	1	0	0	0	0
Bluntnose minnow	Omnivore	49	60	2	3	738
Johnny darter	Insectivore	21	6	1	26	63
Fantail darter	Insectivore	164	280	88	3	167
Rainbow darter	Insectivore	10	193	0	2	265
Greenside darter	Insectivore	10	206	0	0	92
Northern hog sucker	Insectivore	44	35	0	1	126
Silverjaw minnow	Insectivore	15	0	0	2	175

Striped shiner	Insectivore	44	35	0	10	257
Blacknose dace	Generalist	47	8	16	4	0
Stonecat madtom	Insectivore	1	10	0	0	0
Rock bass	Carnivore	0	4	0	0	1
Golden redhorse	Insectivore	0	1	0	0	1
Green sunfish	Insectivore	0	0	0	1	6
Redside dace	Insectivore	0	0	0	1	0
Sand darter	Insectivore	0	0	0	0	86
Smallmouth bass	Carnivore	0	0	0	0	5